TOMATO GROWING TODAY

TOMATO GROWING TODAY

Ian G. Walls

David & Charles : Newton Abbot

This book is dedicated to many friends
who grow tomatoes for profit or fun, and
who have taught me much about successful
tomato growing

ISBN 0 7153 5435 6

© Ian G. Walls 1972
Second edition, revised 1977

Set in eleven on thirteen point Bembo
and printed in Great Britain
by Redwood Burn Limited, Trowbridge & Esher
for David & Charles Publishers Limited
Brunel House   Newton Abbot   Devon

Contents

Chapter   Page

    LIST OF ILLUSTRATIONS   7

1   SETTING THE SCENE   9

2   SELECTION AND SITING OF GREENHOUSE   11
     Basic requirements for tomato culture – greenhouse erection

3   GREENHOUSE EQUIPMENT   28
     Heating and ventilation systems – equipment for watering and feeding – benches, growing rooms, supplementary lighting, lining

4   GROWTH AND NUTRITIONAL NEEDS   47
     Physiological aspects of growth – plant nutrients and the tomato – tomato base fertiliser formulations

5   NUTRIENTS AND GROWING MEDIA: SPECIFICATIONS   69
     Soil and tissue analysis – estimating nutrient requirements – basic ingredients for growing media – compost chart

6   TOMATO PROPAGATION   87
     Germination – programming the crop – sowing and potting – care of young plants – guide to young plant raising – grafting

7   AN EXAMINATION OF CULTURAL METHODS   104
     Growing methods 'balance sheet' – aspects of water application

8   PRE-PLANTING PROCEDURES   113
     Selection of cultural method – preliminary considerations for border cultivation – initial preparations for grafted

tomatoes – for ring culture – for growing in limited quantities of media (bag, box, trough, mattress, pot, bucket) – for straw bale culture – single and double truss cropping

9 PLANTING PROCEDURES 136
Planting distances – the right time to plant – planting procedure in borders – modifications for other cultural systems

10 ESTABLISHMENT AND TRAINING PROCEDURES 145
Early treatment – watering – support – methods of training – pruning – stopping

11 SEASONAL GROWING PROCEDURES 158
Growth factors requiring attention – seasonal feeding programme – foliar feeding – mulching – cultural modifications for other systems – fruit picking to end of season

12 DISEASES, PESTS, PHYSIOLOGICAL AND NUTRITIONAL DISORDERS 173
Damping-off – root disorders – wilts – diseases affecting stems, leaves and fruit – virus disease – pests – pests and diseases chart – mineral disorders – physiological troubles

13 STERILISATION OF SOIL 200
By heat – by chemicals – chart

14 TOMATO VARIETIES 207
General classification – variety list – notes on list

15 OUTDOOR TOMATOES 220
Pre-planting procedures – planting – training and general culture

APPENDICES

1 PEST AND DISEASE CONTROL IN TOMATOES 226

2 NOTES ON TOMATO GRADING 228

3 METRIC AND OTHER CONVERSION FACTORS 230

ACKNOWLEDGEMENTS 232

INDEX 233

List of Illustrations

PLATES

	Page
Dutch light glasshouse (*Scottish Field*)	49
A wide-span glasshouse designed for maximum light catchment (*Scottish Field*)	49
Bubble greenhouse	50
Earlier type of plastic greenhouse	50
Warm-air oil-fired heater for large greenhouse	67
Fan ventilation	67
Aspirated screen for temperature control	67
Sprayline for water and liquid feed	68
Trickle irrigation (*Scottish Field*)	68
Ring culture (*Scottish Field*)	68
Lower propagation temperatures: good bottom truss (*Scottish Field*)	149
Higher propagation temperatures: small but early lower trusses	149
Straw bale culture (*Scottish Field*)	150
Typical root system in straw bale culture (*Scottish Field*)	150
Tomato grafting (*Scottish Field*)	167
Author's greenhouse: heavy crop on chemically sterilised soil (*Scottish Field*)	168
Rotary drum steriliser	168

7

FIGURES IN TEXT

		Page
1	Climatic aspects of British tomato production	12
2	Base walls restrict winter entry of light	16
3	Evenly pitched roof: winter and summer	18
4	East-west greenhouse with roof of uneven span	19
5	Catchment of water from higher land while maintaining level border	23
6	A mobile greenhouse allows flexibility of cropping	24
7	Wind protection without sun exclusion	26
8	Scale plan of greenhouse	30
9	Pipe layout	33
10	Main aspects of environmental control	37
11	Electrically warmed bench suitable for tomato propagation	43
12	Section through a growing room (Electricity Council)	44
13	Capillary bench	45
14	Tomato flower—main parts	53
15	Diagram of tomato plant showing functions	54
16	The bushel—comparison of quantities	85
17	The Jones Rothwell evaporimeter	110
18	Ring culture	125
19	Polythene bag culture	128
20	Trough culture	129
21	Straw bale culture	133
22	Spacing of plants	137
23	Soil-warming pipes	139
24	Plant training methods	154

1

Setting the Scene

I think it can truthfully be said that most gardeners and growers realise that the tomato is not a native of this country, but of lands much warmer than Britain. The original *Lycopersicon esculentum* is a small-fruited plant native to South America, and while I have never been there I should imagine that it sprawls around the countryside in certain areas as a weed, in much the same way as the weeds of our own hedgerows and fields in Britain.

There is of course nothing unique in the cultivation of a non-native plant, and one only needs to carry out a little research on many of our common garden plants for confirmation of this. There are many countries where the climate is very similar to our own in respect of winter and summer temperatures, light intensity, rainfall, and so on, yet the region of South America which forms the natural habitat of the lycopersicon is not one of them. The lycopersicon for one thing, is not a hardy plant and will stand no frost, and this applies to all vegetative portions of its anatomy, including the roots. To thrive well, which means not only vegetative development but successful production of seeds and the ripe flesh which surrounds them, tomatoes require warmth and a fairly high light intensity, as well as a sufficiently long season of growth; this rules out Britain and some of the USA for outdoor culture, except for a brief summer period in favoured areas.

The tomato as we know it today is, of course, very different in physical characteristics from the small-fruited lycopersicon, and this is a result of a controlled breeding programme over many years and the endeavour to introduce desirable features for different purposes.

It is interesting to see that modern breeding has still not altered the basic growing habit of the lycopersicon, which is to produce a many-branched plant intended to sprawl over the ground. The intensified culture, part and parcel of the modern tomato variety, contrives to contain growth in a form suitable either for a greenhouse or for a frame, or for beds out of doors. In most of Europe the tomato is treated largely as a glasshouse crop, but as one approaches the Mediterranean zone culture out of doors becomes increasingly more successful.

The type of tomato grown varies considerably in different countries; in most of Central Europe the relatively small-fruited tomato is in demand, but the pattern changes to large-fruited fleshy tomatoes in the warmer climates, and this includes the larger part of America.

The intelligent understanding of all the inherent wants of the tomato adds up quite simply to successful culture, and it is this we will now proceed to discuss in detail.

Obviously there are bound to be many different views on the 'right' way to grow tomatoes, especially if one considers the whole vast sphere of commercial, professional and amateur culture. What I have tried to do in this book is to discuss the whole area of tomato culture in a logical and systematic manner, taking into account the views of growers and research workers in this and other countries. While some of the aspects dealt with are highly technical, I have tried to reduce matters to practical terms, so that everyone who grows tomatoes can benefit in some way from the information which is given.

2

Selection and Siting of Greenhouse

BASIC REQUIREMENTS FOR TOMATO CULTURE

Climate

Many countries, even relatively small ones, offer a considerable variation in outside temperatures over both their length and their breadth. Generally speaking, for instance, the south of Britain has a higher *average* temperature than the north; and the west is, on average, warmer than the east. Coastal areas enjoy a more equable climate throughout the year, with less frost, due to the thermal warmth derived from the sea, this being particularly true of areas such as western Scotland which enjoy the full benefit of the Mid-Atlantic Drift, commonly called the Gulf Stream. Exposure to wind can, however, nullify much of the benefit of latitudinal or coastal placement in several ways: apart from an overall reduction in temperature out of doors, heat loss is increased in any glass or polythene structure; physical damage to plants growing out of doors is also highly likely.

Rainfall patterns in most countries are extremely variable, according not only to latitudinal placement, but to precise location in relation to hills. Again as an instance, for the most part, eastern Britain is drier than the west, and likewise northern Britain is wetter than the south, though the dryness of the east persists fairly far north. Relating such weather patterns to tomato culture, it can be said that tomatoes can be grown relatively easily anywhere in Britain or, say, cooler areas of the USA, with the protection of a greenhouse or other covering device (cloches, elevated frames, etc); whereas successful outdoor culture (the plants having been raised under glass) is only possible in the warmer parts of the country and even there real success is dependent on a good summer.

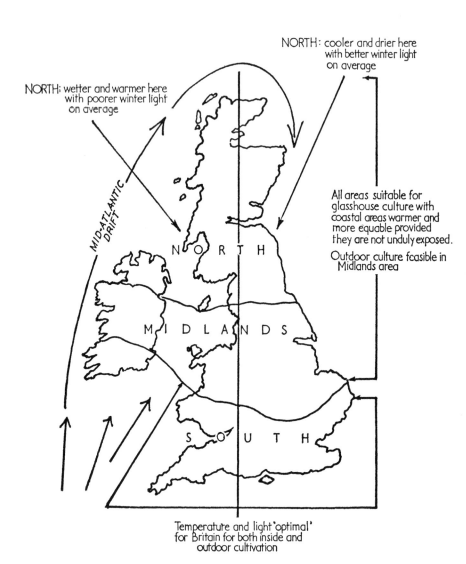

NORTH: cooler and drier here
with better winter light
on average

NORTH: wetter and warmer here
with poorer winter light
on average

MID-ATLANTIC DRIFT

NORTH

All areas suitable for
glasshouse culture with
coastal areas warmer and
more equable provided
they are not unduly exposed.

Outdoor culture feasible in
Midlands area

MIDLANDS

SOUTH

Temperature and light 'optimal'
for Britain for both inside and
outdoor cultivation

1   *Climatic aspects of British tomato production*

Heat is certainly necessary for the propagation stage of tomatoes, no matter how they are later cultivated; the exception is a really late crop which can be propagated without artificial heat. Even then, propagation heat is more or less essential if any timetable is to be successfully adhered to.

Culture in greenhouses can range from completely cold to relatively hot, but there is a great benefit in having artificial heat to avoid high humidity problems and delayed ripening, especially in the northern parts of the country where a cold greenhouse in a poor summer can result in a very late and much reduced crop. The actual period of culture in greenhouses will depend not only on the level of heat provided, but on light intensity, as good light is essential to sustain photosynthesis in the leaves and generally satisfy the tomatoes' metabolic system (see below).

Light Intensity

Temperature patterns are fairly closely related to light intensity, not only to the total 'bright sunshine' registered by meteorological stations, but to total radiation, which is bright sunshine plus *diffused* light. In other words, an area with higher bright-sunshine figures also enjoys higher total radiation in consequence. The vital period for light intensity is the winter, when in northern climes the available natural light falls well below the requirements of the tomato plant, for vegetative development and the production of fertile pollen followed by successful fertilisation. It is interesting to note that where there is a large bulk of water, either coastal or inland, total radiation figures are higher, as reflected light is added to the direct light. An area free of smoke pollution and not subject to fog will also enjoy higher total radiation figures, not only because of the unrestricted passage of light through the atmosphere, but because there is less staining of glass or polythene. Sun shut-off by hills, buildings or trees must also be taken into account in any area.

Nutrition

On the growing tomato plant there must be roots capable of sustaining foliage, flower and fruit development, and these roots must be sufficiently healthy and unaffected by pest or disease, otherwise the whole growing process is either arrested or slowed down, a situation by no means uncommon (see Chapter 12). Adequate nutrition is also essential, especially for the plant bearing a heavy crop of fruit and swelling this without the assistance of a normal complement of foliage, much of which is frequently removed under modern training methods.

A final point, not perhaps so widely appreciated, is that where artificial growing conditions prevail in the culture of any plant, the possibility of pest or disease attack is far more likely. A greenhouse is entirely artificial, as there is no natural rainfall and all water and nutrients have to be supplied, in effect, artificially by the gardener or grower. Humidity and temperature levels also lie largely in the hands of the operator, as indeed does the choice of growing medium.

Though it would be foolish to imply that there was anything particularly unique about tomato culture, it is important to appreciate the fact that tomatoes are plants with many whims to satisfy.

GREENHOUSE ERECTION

There must be compromise in most things, and this certainly applies to the choice and siting of a greenhouse for the growing of tomatoes. It would be simple to list all the requirements: the selection of a good light area, a flat site, good drainage, adequate water and power supplies; the choice of a design allowing for maximum light transmission, and the installation of automatic heating, watering, ventilating and feeding systems. Such a procedure is frequently followed in articles and books relating to the commercial production of tomatoes, and the points they make are indeed valid.

Yet to most gardeners a garden, yard or greenhouse is an adjunct to the dwellinghouse, which is selected on many considerations, often mainly congenial living conditions and proximity to place of business. It is true, of course, that enthusiastic gardeners may be strongly influenced in their choice of dwellinghouse or site by the horticultural

potentialities of the garden. Indeed one wonders just how many of the older and larger estate-type dwellings are literally developed around the garden, and not merely because of its production capacity, but because of its landscape potential.

To be thoroughly practical, however, one must admit that normally a greenhouse has to become part of an existing garden, though what has been said about climatic patterns should not be overlooked in the selection and siting of it.

Type of Structure

The perplexities of selecting a greenhouse with tomato culture largely in mind are manifold. Light transmission, and this obviously includes solar heat rays, is highly important, and it is advisable to think in terms of an all-glass type of structure where the glass-supporting and retaining components are minimal, the glass itself being the most important constituent. This usually means little or no base wall, whether of brick, block, wood or asbestos. The tomato, though undoubtedly a tall crop, demands light for its full life and especially when newly planted out, and base walls would tend to reduce light transmission to the young plants (fig 2).

I mentioned earlier the need for compromise, and here is where it comes into play, as quite frequently the tomato crop is Cinderella to the over-wintering of pot plants and dormant tubers, followed by some varied early-season propagation. Such activities demand constant heat and certainly 100 per cent frost protection during the winter months. All forms of artificial heat cost money, and as the chief value of glass is its ability to transmit solar light and heat it follows that glass is a poor insulator and does not readily conserve heat when solar radiation ceases.

In effect, therefore, the more glass in any structure, the higher the heating costs, and while the all-glass structure can provide optimal conditions for tomato growing as far as light transmission is concerned, one has to be prepared to pay for the artificial heat necessary to get full benefit of its light-transmissional qualities. This must be linked to the cost of heating, not only for tomatoes, but throughout the winter for the other purposes referred to, and it may be preferable to select a base-wall type of house which will obviously be less costly to heat especially throughout the winter months (see Chapter 3). Where cold growing is intended and the cost of artificial heat of no consequence, then one should certainly choose an all-glass greenhouse.

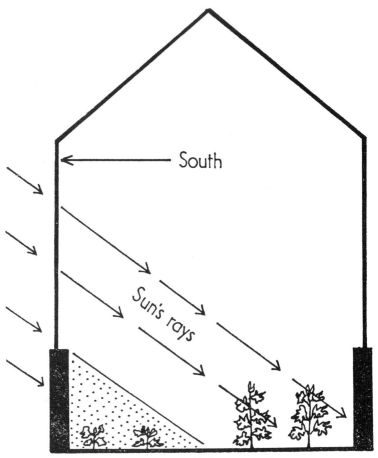

2 *Base walls restrict entry of light to young plants in winter*

It is interesting to note that while glass traps the long solar heat rays reflected back by plants, soil, benches etc, polythene does not, and this is why some polythene structures lose their sun heat more quickly than glass (see page 29).

Orientation

Simply because a greenhouse has a high proportion of glass to opaque material, this does not ensure the best possible light transmission. Glass has reflective qualities and transmits maximum solar light and heat only when the angle of incidence is normal or nearly so (90°). The angle of the sun in the sky does, as we all know, vary considerably, not only between morning and evening but also according to season. In the winter, as compared to summer months, the sun is in the sky for very brief periods and the low angle of incidence of its rays on the earth's surface is such that heat and light rays are deflected from all but the nearly vertical sides or ends of a greenhouse. Much of the solar radiation is certainly deflected from the normally angled roof of a conventionally shaped greenhouse. It may be difficult to appreciate this phenomenon simply by using the eye as a measuring device, yet light-sensitive plants will certainly register their protest by showing drawn or etiolated growth, a frequent state of affairs with light-loving house plants in the home. The light-transmissional qualities of a greenhouse can, of course, be measured by instruments such as photo-electric cells.

It follows, therefore, that when winter light is particularly desirable, as it is for young tomatoes or for most other propagation activities, it is important to site a greenhouse with its long axis running east-west so that the maximum area of glass is presented favourably to the low-angled winter sun (fig 3). As the sun rises on the horizon the south-facing roof slope gradually absorbs most of the solar light and heat, but note that any obstruction to the sun on the south side of the greenhouse, such as densely planted tomatoes when they get larger or other tall crops, will shade plants at the rear of the greenhouse. With the average-sized amateur greenhouse, however, this is not usually a major consideration. If a greenhouse is sited with the ridge running north-south, only the south-facing gabled end is suitably angled to absorb winter light. When however the sun is higher on the horizon, from spring to autumn, there is more equal light distribution from the east in the morning and from the west in the afternoon and evening, and this has its advantages. On balance, however, there is much to be said for an east-west siting for the single greenhouse.

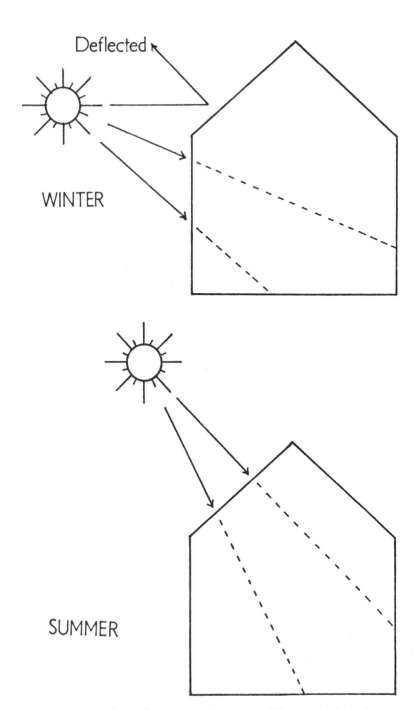

3 *An even-pitch greenhouse orientated east–west deflects much of the low-angled winter sun from its roof; the situation changes in summer when the sun is higher*

Shape of Greenhouse

A considerable amount of research has taken place over many years on the best configuration of greenhouse for maximum light transmission, winter and summer. In commercial circles unevenly angled houses (see fig 4) with steeper south-sloping roofs have been experimented with, as also have mansard-shaped greenhouses. In amateur spheres, circular-type houses have now been on the market for some years, some in fact capable of turning to allow plants to catch the sun, but I cannot feel that these circular houses have a great part to play in tomato culture, other than for the propagation period when they would certainly be extremely valuable. Mansard-type greenhouses are also readily available in amateur sizes.

It is not perhaps too well known that there is usually more than

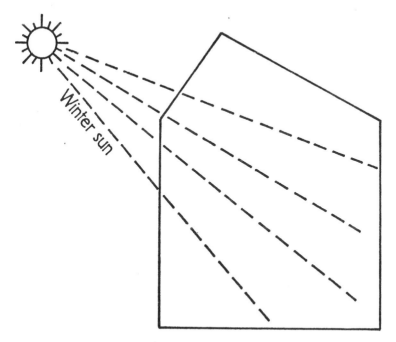

4 *An uneven-span greenhouse orientated east-west with a suitable pitch to south absorbs more winter sunlight*

adequate light for tomatoes during the summer months, the really vital periods being during winter, spring and autumn; more will be said about this later. Sun shut-off occasioned by hills, buildings or trees should be looked out for. I have frequently come across excellent greenhouses situated in a particular location in the garden where a tree shades off the sun for several hours each day—although this is not necessarily a great disadvantage in high summer.

Provision of Shelter

Exposure to wind is a matter of considerable importance, as it can affect the heat-loving tomato plant in many ways, especially by reducing the average temperature in the greenhouse by accentuating heat loss; it also raises the fuel bill. Periods of low temperature resulting from failure to take exposure into account impose an inevitable growth check which can be very damaging to the tomato crop (see Chapter 4). Outdoor culture would, of course, be similarly affected by exposure.

Provision of shelter is often therefore a necessity, either by the use of artificial screening material or by the planting of hedges or tree belts—provided these are at a sufficient distance to avoid sun shut-off or root problems. It is now accepted that a 50 per cent permeable shelter material is more efficient than a solid one, as it does not create turbulence. A screen or hedge should provide effective shelter over a distance of approximately 10–15 times its height.

SHELTER MEDIA—Artificial

Plastic trellis	— Clean, long-lasting
Wood trellis	— shorter term, cheaper
Hollow blocks	— great possibilities
Netlon	— portable, very effective
Horizontal cane	— quite effective
Vertical canes	— quite effective
Alternative offset slats	— offer privacy also
Wire netting and straw	— effective, but treat with flat black paint

Vertical lath	— must be treated
Single layer slats	— very impressive looking
Interwoven fencing	— needs heavy support if exposed
Polypropylene	— rot resistant
Rokoline	— effective and attractive
Horizontal lath	— must be treated

SHELTER MEDIA—Natural

Hedges

Myrobalan (*Prunus cerasifera*)	⎫
Prunus pissardii nigra	Make thick
Holly (common form)	impenetrable
Yew	hedges up
Privet	to 6–8ft;
Beech	require
Hornbeam	regular
Quickthorn	cutting
Willow (kept as hedge)	⎭

Trees

Silver Birch; Larch;	Make effective
Picea excelsa and *sitchensis*;	shelter belts
Poplars, Italian and *canadensis*;	if sufficient
Limes; Mountain Ash;	room for
Pinus sylvestris	development

Conifer hedges

Chamaecyparis lawsoniana	Make tall hedges
Cupressocyparis leylandii	up to 20ft
Thuja lobbii and *excelsa*	or more

Further Aspects of Greenhouse Selection and Siting

Other issues of some importance regarding greenhouse selection are width of doors to allow ready access, sufficient headroom, constructional materials and glazing systems of such a type to avoid constant maintenance—this I feel being of paramount importance. Ventilation must be adequate to avoid over-heating and excess humidity. Vents should preferably be on both sides of the ridge. It can be achieved either by lifting vents or fans (see page 35) to allow a sufficient number of air changes. Warping or distortion are also important points to bear in mind. It is for this reason that aluminium alloy greenhouses are enjoying such popularity, although I see nothing wrong with a well-constructed wooden house, provided it incorporates a grooved glazing system and that superior or properly pressure-treated wood is used. Other metal-type houses also have distinct virtues, provided the metal is well treated to avoid corrosion.

STRUCTURAL MATERIALS

Pine	— difficult to treat with preservative
Redwood	— good if painted or treated
Red cedar	— good but not strong
Hardwoods	— good but rather expensive
Burma teak	— very good, also expensive
Pressure-treated softwoods	— cheap and fairly good if well treated
Steel	— must be galvanised or otherwise treated
Cast iron	— less liable to rust than steel
Reinforced concrete	— good, but little used
Aluminium alloy	— good in every respect

Erection on Site

Always follow erection plans carefully. If levelling a site by one or other of the methods available (cut-and-fill, or levelling to highest or lowest point) always allow a period for subsidence. Ensure that there is adequate drainage, especially on a sloping site where water from higher land could be a severe problem for border tomato growing. It is also highly important that the subsoil should not be unduly consolidated during levelling procedure, otherwise drainage could be further affected (see fig 5, showing greenhouse erected on sloping site with appropriate drainage installed).

Control of bad weeds in the erection area must be carefully carried out, remembering that any chemical used for control purposes will not readily wash out of the greenhouse border once the area is covered by the greenhouse. It is therefore advisable to treat the bed some time prior to erection (see page 117).

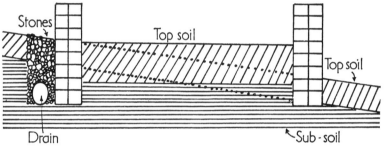

5   *A method of catching water shed from higher land, yet providing a level border*

Adequate supplies of water and electricity are essential. There should also be collection or drainage for water shed from roofs where this is likely to be a problem.

Mobile Greenhouses

Any account of tomato culture would be incomplete without a reference to greenhouse structures that can easily be moved to allow a crop rotation system and thereby avoid the problems of monoculture—for soil sickness invariably results from cropping continuously with tomatoes on the same site.

Simple Dutch light greenhouses on temporary base block 'walls', or complete light sectional greenhouses of any type, can of course be readily moved around the garden. This is perhaps even more true of plastic structures, particularly as the plastic requires fairly regular renewal in any case.

For a period after the last war there was a great vogue in commercial-size greenhouses capable of movement on rails or wheels inset on 'dollies', and mobile houses of all types are still available if specially requested. It is interesting to note that mobile greenhouses were first used in Britain during the early part of the century, not only for tomatoes but for tobacco. The only type likely to interest the amateur, however, is a Dutch light structure with pulley wheels incorporated and running on rails set in the base blocks.

Any type of mobile structure does, of course, allow crop rotation by permitting the house to be moved *en bloc* over a predetermined number of plots, with obvious advantages where border culture is concerned (see fig 6). But higher initial cost of structure, plus the problem of a practicable form of heating and provision of other services, are disadvantages which must be carefully considered.

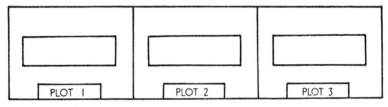

6 *A mobile greenhouse allows considerable flexibility of cropping: plot 1, lettuce followed by tomatoes, moving to plot 2 for autumn chrysanthemums planted out in May, moving to plot 3 to cover crop of winter lettuce sown in August. For the second year start the procedure in plot 2*

The efficiency of modern soil sterilants, coupled with the increasing number of alternative cultural methods, have tended to make mobile houses much less attractive.

Plastic Structures

Various types of plastic structure can of course be used for tomato culture with considerable success. There are two basic forms available. The first is a framework of wood or metal over which the plastic is stretched, being kept in place either by nylon rope, nylon netting, spring tensioning, battens of wood, or by simply being bedded into the soil. Ideally there should be an avoidance of nails to reduce strain, and care should be taken to prevent the plastic lying too long against a solid wooden or metal supporting member, for this allows heat to build up and damage the plastic. The other type of plastic structure is the bubble house, which is kept inflated with a fan and has a counter-balanced ancillary fan for ventilation. This form has attracted attention in recent years, and it too can be used for tomatoes.

Plastic, however, does deteriorate. Absorption of ultra-violet rays can result in its chemical breakdown, and indeed is the main problem with most forms of plastic, although ultra-violet proofed materials are now available and are reasonably successful. Deterioration also results from dust adhesion caused by static electricity, and it is therefore generally advisable to renew the plastic every two years or so, although there are no hard and fast rules about this. Furthermore, all the light admitted by plastic is diffused, *not* direct, and unlike glass it does not trap the reflected long rays, which in effect means that there is more rapid cooling at night, giving rise to condensation—ventilation methods for plastic greenhouses always require careful consideration. Special provisions for crop support may also be necessary in certain types of plastic structure, but rigid sheets of plastic, either flat or curved, are of course now readily available and can be used in conventional greenhouses instead of glass, though they generally tend to be expensive.

GREENHOUSE ERECTION SUMMARY

Sites: open—avoidance of sun shut-off for prolonged periods. Where exposure is a problem suitable shelter should be arranged, taking into account prevailing winds and avoiding excessive sun shut-off (see fig 7).

Greenhouses: all-glass types ideal for both propagation and culture due to maximum light transmission winter and summer, although more costly to heat during winter-spring period when solar radiation is low. Base-wall-type houses are less costly to heat but afford poorer light transmission, although when used largely for winter-spring propagation followed by later tomato crops, this is not a great disadvantage. This is especially true of the summer period, when excess of light and solar heat can be a problem.

Ventilation: conventional ventilation area (if capable of being fully opened) at least one-fifth (20%) of floor area. Preferably operated by

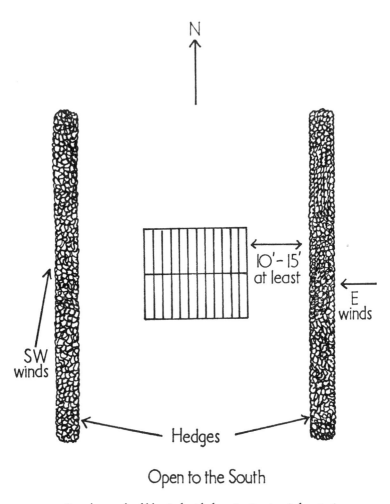

7   *Greenhouses should be sited with due attention to wind protection
and the avoidance of sun exclusion*

expansion-type vents or electric automatic systems. Fan ventilation,
either extraction or pressurised, should be designed correctly with
adequate capacity fans and adequate inlet/outlet vents (see Chapter 3).
Ease of maintenance: aluminium alloy, galvanised or enamelled
steel, superior woods or pressure-treated softwoods. Glazing methods
should be modern, avoiding the use of putty.

Design: easy access and adequate headroom. Sloping sides will usually result in some lack of headroom and crop height.

Erection: carried out following careful levelling and weed control. Time to be allowed for subsidence. Essential to bring to site a good supply of water of suitable pressure and electricity of sufficient loading (see page 31).

3

Greenhouse Equipment

HEATING SYSTEMS FOR TOMATO GROWING

Much could be written about the installation of heating systems, which is a highly specialised subject. For tomato growing three main facets must be carefully considered: (1) adequate heat level, (2) good distribution of heat, (3) precise and preferably automatic control of greenhouse heat.

Adequate Heat Level

To calculate the heat input necessary for any building, including greenhouses, the procedure is much the same. All structural materials allow the passage of heat or cold in either direction (this is called the U-value) and the rate at which this occurs depends on a number of factors. Glass is obviously efficient in transmitting solar heat, which makes it a poor insulator. Opposite (top) are the important figures to remember, bearing in mind that they refer to U-value in British thermal units per square feet per hour. For the metric equivalent per degree C multiply by 5·7 (see Appendix 3).

To calculate the heat loss of any greenhouse (or other building):

1. Draw a scale plan of the greenhouse and append to this accurate measurements of the glass areas and of the base walls of brick or wood, where these are involved (see fig 8).
2. By reference to the Btu table calculate the heat loss per degree of temperature difference between outside and inside temperatures.
3. Now multiply by the temperature 'lift' you think will be necessary over outside temperatures. For frost protection or late tomato crops it is usual to multiply by 20, assuming that if it is 30° F out of doors,

Material	Btu/hour per sq ft per degree F
Glass, including framework	1·1
4½in brick wall or composition block	0·5
Double brick wall, 9in	0·4
Wood 1in thick	0·5
Asbestos (sheet or corrugated)	1·1
Concrete 4in thick	0·75
Double glazed glass (properly sealed)	0·5
Polythene all gauges (approx)	1·1

Note: higher figures, eg glass 1·4, single brick 0·64, are sometimes quoted especially for older greenhouses. In practice, and with tightly glazed modern greenhouses, the addition of one-third to the final calculation allows for the lower figures to be taken initially.

Sides and ends* (base wall)	$2 \times 2' \times 10' + 2 \times 6 \times 2'$ (ends) $= 64$ sq ft heat loss of 0′5 Btu $= 32$ Btu
Sides (glass)	$2 \times 4' \times 10' = 80$ sq ft ⎫ = 210 sq ft at
Ends* (glass)	$2 \times 6' \times 4' = 48$ sq ft ⎪ heat loss of
Roof	$2 \times 3' 6'' \times 10' = 70$ sq ft ⎬ 1·1 Btu =
Gable ends (glass)	$2 \times 3' \times 2' = 12$ sq ft ⎭ 231 Btu
Total heat loss =	$231 + 32 = 263$ Btu per hour
Allowing for a 20°F lift	$263 \times 20 = 5,260$ Btu

* Door included for easier calculation, but can be allowed for if desired.

50° F will be maintained in the greenhouse (see table below); eg, $263 \times 20 = 5,260$ Btu; for early crops or a minimum of 60° F multiply 263 by 40 = 10,520 Btu.

4. Now add a factor which takes into account how well constructed

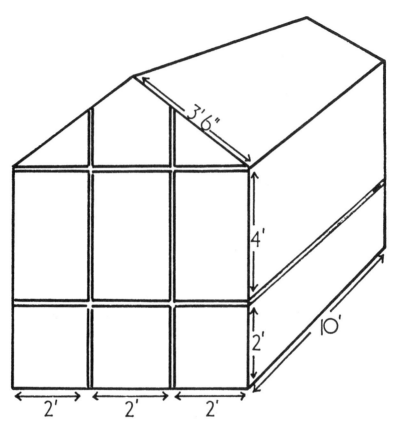

8   *Scale plan of greenhouse*

your greenhouse is, as a leaky structure with badly fitting doors and glass loses heat more quickly than a well-constructed tight structure. A greenhouse in an exposed situation will also suffer greater heat loss than one in a sheltered position. Normally for a modern structure one-third is added to cover this, eg 5,260 + $\frac{1}{3}$ (1,753) = 7,013Btu. This is the heat loss of the greenhouse for the heat lift decided upon, and it is now necessary to select a heating system capable of giving this number of Btu, at the same time relating this to running costs.

A 2$\frac{1}{2}$kW electric heater has an output of 2$\frac{1}{2}$ times 3·412 (the Btu value of 1kW) = 8,530Btu, and this would therefore be adequate for

moderate heat and frost protection of the greenhouse we have calculated for. The cost of running it can readily be found by relating this to
the respective unit cost, normally taking about a 50–60 per cent demand,
except during very cold weather. This means that to run a thermostatically controlled 2½kW heater for 24 hours the total consumption
will be 50 per cent of 24 × 2½ (60) = 30. 1kW is the unit used for
costing purposes so the running cost per 24 hours would be 30 units at
the appropriate tariff. A much more realistic attitude to frost protection
where electricity is concerned is to set the thermostat incorporated in
most electrical heaters to come on at 43–45° F and this can reduce the
demand very considerably according to the weather pattern, although
for tomatoes higher temperatures are desirable. Where hot water pipes
are used, the calculated heat-loss figure is divided by the Btu output per
foot of the chosen pipe diameter and this gives the length of piping
required. Gilled pipes give up to five times the output of plain pipes
(according to design).

Pipe diameter	Btu (to nearest round figure) at normal operating temperature (150° F) with greenhouse at 55° F
1in	70
1¼in	100
1½in	110
2in	130
2½in	150
3in	190
3½in	200
4in	230

Note: As temperature of pipes increases, heat emission increases
accordingly. For example, for the calculated heat loss of 7,013 and
using 3½in diameter piping, the length of piping required will be
$\frac{7,013}{200}$ = 35ft. Regarding running costs, if an oil-fired boiler is used the
calorific value of a gallon of fuel oil is taken at 160,000Btu less 25 per

cent loss on ignition = 120,000Btu; therefore for a heat loss of 7.013Btu a gallon of oil costing approximately 25p will give $\frac{120,000}{7,013}$ = 17 hours approximately. If the norm of a 50 per cent demand is taken, this becomes 34 hours. For solid fuel the same procedure is followed.

These figures are based merely on frost protection, but a decision must be made about the temperature level necessary to ensure that the *air* temperature in the greenhouse cannot drop below an acceptable level, which for tomatoes is around 55° F (13° C). An important consideration now is to take into account at which period of year the maximum temperatures will be required. If, for example, winter propagation followed by tomato planting in February is desired, due allowance must be made for coping with outside temperatures of a very low order (possibly 15° F − 9° C). Taking this into consideration it can be seen that a temperature 'lift' of about 40° F over outside temperatures must be allowed. This means multiplying the heat-loss figure by 40. For tomatoes planted in April or May even a 10° F lift may be acceptable, much of course depending on the region of Britain, exposure, etc.

With cold culture no heating is of course involved, although experience shows that in the northern part of Britain completely cold culture of tomatoes can result in a very late crop and raises problems of excess humidity—with its disease implications.

Good Distribution of Heat

Heat is provided by radiation and by convection, and invariably a combination of both. Radiation heat, which is directional, will be given off by warm pipes, but convection currents will also be developed as the air warmed by radiation in the vicinity of the pipes rises, setting up convection air movement.

Non-fan convector electric heaters operate in a similar fashion by warming air which discharges itself and sets up convection currents, whereas fan-type heaters blow the warm air out in a specific direction, but here again convection currents are also developed. Free-discharge heat from a central source in a greenhouse is not as efficient as heat properly distributed, simply because cold curtains of air quickly develop due to the rapid heat loss through the glass. Different air movement on the outside of the greenhouse can also encourage the development of these cold areas. Warm-air heating systems can only be considered efficient when the air is distributed through polythene ducts, but it should be explained that such efficiency is only really of importance for early crops when outside temperatures are likely to be low.

So far we have only considered warming the air, but of course it is highly important for tomato culture that the soil or growing medium is sufficiently warm—not below 56–57° F (about 13° C) at 4–6in depths. It can be seen therefore that there are considerable difficulties in achieving this soil warmth uniformly when relying entirely on non-directional warm-air systems, and the same may be said of badly designed pipe systems.

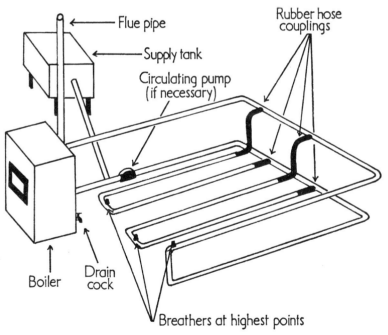

9 *Pipe layout for small tomato house*

The ideal pipe system for tomato growing is one designed on the small bore principle, using 1¼–2in diameter pipes. These small diameter pipes, containing as they do only one-fifth of the volume of water of an old-fashioned 4in system of similar heat capacity, can be readily spread across the growing area between the plants, and on the soil if necessary, which they warm by radiation heat. In addition they are, by virtue of the smaller volume of water, able to respond to the need for heat much more quickly than a 4in system with its large bulk of water, and this is important for tomato growing where precise temperature control is desirable. The ideal layout of pipes for a single span of a small tomato house is shown in fig 9; note also that if rubber hose combined with expanding clips is used for the joints of the interspersed pipes, this allows them to be left on the soil for soil warming by radiation and conduction early in the year and lifted up a few inches thereafter to give convection heat. Where long lengths of small-bore pipes are used a circulating pump is necessary, and here lies the main advantage of large-bore systems (3–4in), in which the water circulates quite readily by gravity, provided there is a gradual rise from the boiler to the highest point and a gradual fall back thereafter.

Heating Methods and Costs

Types of heating system for tomatoes or greenhouses generally need not of course rely on small- or large-bore water-filled pipe systems. The following is a summary of heating methods:

1. **Water-containing pipes**—boiler fired with solid fuel, oil, gas or electricity. Simple types are not automatic. More refined types are semi-automatic. Refined oil, gas or electric boilers are completely automatic.

2. **Oil heaters**—free discharge or ducted heated air. These can be with or without pressure jet, vapouriser burners, or assisted air discharge by means of fans. Simple types are controlled manually, other types operate on a thermostat.

3. **Various forms of electrical apparatus**—all reasonably automatic.
    a. Soil- or bench-warming cables (see page 43), which have little effect on air temperature.

b. Mineral-insulated (MI) cables fitted around the greenhouse perimeter.

c. Tubular heaters, generally fitted on perimeter walls.

d. Fan heaters, which are generally free standing in the centre of the greenhouse, but there is a larger type which can be sited at one end of the house.

e. Storage heaters—there are some problems with these owing to lack of proper temperature control.

The **approximate** cost of operating the various heating systems can be calculated from the table shown on the next page.

It is obvious that for full-level heating, solid-fuel and oil-fired systems are much more economic to run than electrical systems, but on the other hand they are generally much more expensive to install, and they frequently offer considerable control problems compared with electrical heating units which are simple to control thermostatically, a matter which will be discussed in the next page or so.

Ventilation

While much has been said about the tomato plant's need for warmth, it must also be stressed that tomatoes dislike excess heat and very high humidity. Excess heat, apart from causing wilting or scorching, can lower fruit quality. Reducing the temperature of a greenhouse when either solar heat or artificial heat (generally residual), or both, raise the temperature above a desirable level can be effected in three ways: (1) The warm or lighter air is allowed to escape through suitably large ventilators set in the highest part of the structure, and is replaced by cooler outside air by reciprocal interchange or inadvertently by entry through leaks in glass overlaps or joins, the edges of doors, or purposely by lower-set vents. The correct ventilator size, when the vents are capable of opening fully, is one-fifth of the floor area of the greenhouse. The ventilators can be operated manually or by electrical systems which are thermostatically controlled, or better still, they can be expansion-type vents which operate at pre-set temperatures. (2) Extractor-type fans extract the warm air and pull in cooler air through inlets, vents or louvres of suitable size. The fans operate thermostatically and, if neces-

Type of system	Average operational efficiency	Approx running cost per 100,000 Btu (therm) at fuel prices as stated
Simple solid-fuel boiler (hard coal)	50%	Based on coal at 12,500Btu/lb Cost of fuel/ cwt: 120p = 15 p/therm 115p = 14½p „ 110p = 14 p „ 105p = 13½p „ 100p = 13 p „ 95p = 11½p „
Refined solid-fuel boiler (smokeless fuel or coke)	60%	Based on value of 12,000Btu/lb Cost of fuel/ cwt: 120p = 14½p/therm 115p = 14 p „ 110p = 13½p „ 105p = 13 p „ 100p = 12½p „
	70%	120p = 13½p „ 115p = 13 p „ 110p = 12½p „ 105p = 12 p „
Purpose-made oil-fired boiler and fan heaters	75%	Based on oil at 162,000Btu/gal Cost of fuel per gal: 27p = 20 p/therm (bulk purchase 25p = 19 p „ —small quan- 22p = 18 p „ tities approx 20p = 17 p „ 25% extra)
Converted oil-fired with vaporiser-type burner (costs fairly similar for free-standing oil heaters, perhaps slightly more advantageous)	70%	Based on oil at 158,000Btu/gal Cost of fuel: 27p = 20½p/therm (remarks 25p = 20 p „ above 22p = 18½p „ apply) 20p = 17½p „
Purpose-made gas-fired boiler	75–80%	Based on cost of gas per therm. Varies according to district. Add 20–25% to price for loss of efficiency
All electrical heaters (storage heaters not considered here)	100%	Based on normal tariff Cost per unit (3,412Btu): 1½p = 43p/therm 1 p = 29p „ ½p = 14½p „

sary, the inlets too can be operated automatically—but this is generally done manually in small greenhouses. (3) 'Pressurised' systems are installed whereby fans push air into the house, and the warmer air escapes through counterbalanced louvres.

Each system has its virtues and failings. Conventional ventilating systems depend much on the movement of outside air, whereas fan

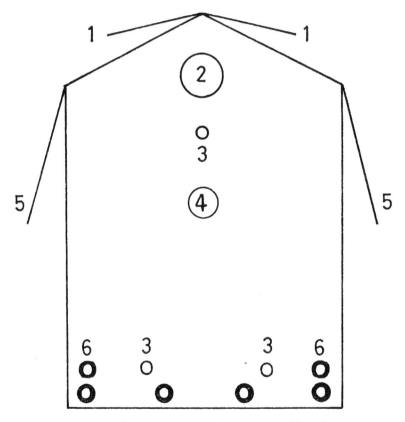

10 *Main aspects of environmental control. Ridge vents (1) preferably automatically operated, in conjunction with side vents (5). Alternatively, extractor fans (2) with properly placed inlets. Aspirated screens (4) containing all thermostats for operation of vents, fans, heating systems, should be situated at crop level. Spraylines (3) at high or low level, or both. Alternatively low-level watering systems can be used. Heating pipes (6) should provide a curtain of warmth in addition to localised heat adjacent to plants*

systems operate more or less independently of outside air movement and are much more positive in achieving the required number of complete air changes (generally 40–50 per hour, or air movement of 7cFm/sq ft), to reduce the temperature of the greenhouse efficiently. Fan ventilation systems do cost money to operate, however, and although the amount of electric power they consume is small, some gardeners and growers seem to grudge it. Breakdown of fans or power failure can be a problem, although where this occurs with the smaller greenhouses the inlet vent coupled with open doors can usually cope for limited periods.

Pressurised systems can result in temperature gradients developing throughout the greenhouse due to localised pressure build-up, but in practice such gradients are small enough to be disregarded.

Controlling Heat Input

There are various methods of controlling the generation of heat in a greenhouse, whether this involves the use of a switch to allow input or cessation of electric power, or the means of igniting or extinguishing a boiler. Obviously the efficiency achieved will vary according to the sophistication of the equipment. For example, a pressure-jet oil-fired boiler can be fully automatic, whereas a vaporising oil burner may only be partly automatic, alternating between the pre-set heat and pilot heat.

A very common way of achieving automatic control with hot water systems where circulation depends completely on a circulating pump is to control the pump operation thermostatically. An alternative to this would be the use of solenoid valves.

Thermostats and other instruments, to be fully effective in a greenhouse, should preferably be aspirated, so that a small fan draws a constant flow of air typical of that in the greenhouse over the controlling instruments. If sited openly all instruments tend to record atypical temperatures or conditions due to the build-up of solar radiation, the effects of draughts, or radiation loss. This applies to thermometers also. Obviously there are limits to the degree of sophistication which can be readily afforded, yet it has been shown that accurate control of heating systems and ventilation results in a much more even temperature regime, with highly advantageous results to crop performance and fuel consumption.

It is possible to achieve almost complete environmental control in greenhouses, including control of temperature and humidity patterns compatible with preceding and prevailing light levels, and such equipment is now relatively commonplace in many modern commercial glasshouse establishments. Further information on glasshouse environmental control equipment is available from specialist suppliers.

WATERING AND FEEDING

In addition to greenhouses and their heating systems, there are certain other requirements for successful tomato culture, particularly when there is a desire to reduce labour by the use of automatic or semi-automatic equipment. Watering and feeding are undoubtedly the two most important issues, and it is difficult to divorce them.

Adequate Water Supply

Much depends on the size of the greenhouse unit in deciding what constitutes an adequate water supply. Large areas of greenhouse require a sufficient volume and pressure of water in order to operate the spray-lines—now an essential part of commercial equipment—and this means that a mains supply pipe of fairly large diameter must be laid directly to the greenhouse. Many water authorities, however, now insist on the installation of a special reservoir tank, usually butyl-lined corrugated iron or something similar, with an electric pump to provide the outflow, and they do this not only in the interests of the grower in question, but because of the liquid fertilisers which may find their way back into the mains supply where there is a direct take-off from the main.

Amateur gardeners, however, should seldom run into these complications, and usually a supply pipe of $\frac{1}{2}$–1in diameter can be installed, either in permanent underground form with water board permission (and possibly an extra charge) or linked to an outside tap and detached when not in use. Such systems need to allow complete drain-off during the winter and are therefore not always suitable for year-round use.

Watering Systems

Sprinkler systems. Sprinklers providing large droplets of water and intended mainly for the damping-down essential to pollination, are fitted at a height of approximately 6–7ft, and whilst in frequent use commercially, are as yet seldom used by amateurs.

Spraylines. These, like sprinklers, are fitted overhead for the early part of the season at about 6–7ft and then inverted and dropped to about 1ft later in the season when the lower fruit trusses are removed. They provide water in fine mist form and are capable of watering the whole growing area. They are not highly suitable for container systems as the water is prevented from entering the container by the foliage of the plant. Good pressure and volume of water are required for the successful operation of spraylines, a matter for precise specification.

Trickle systems. These involve the precise placement of water in droplet form—to avoid soil structure damage—adjacent to the plant and are therefore ideally suited to container growing systems. On certain types of light soil the water tends to form a cylindrical column and restricts root development.

Low-level sprinkler systems. These can involve 'lay-flat' polythene where 2in-wide perforated polythene tubing expands when filled with water and sends out fine jets of moisture. Rigid PVC or plastic hosepipe bored with small holes operate in a similar way. A fair pressure and volume of water are needed for the successful operation of these systems, which are satisfactory for border or trough culture.

Hosepipes. The most common method of watering for the amateur is with a hosepipe, either open ended or fitted with a rose. Damage to soil structure can however, easily occur by continuous and careless use of open-ended hoses, apart from which holding hoses can be a time-consuming method of watering. Damping-down is frequently carried out with a hosepipe, either by pinching in the end of the hose or using a spray nozzle. Hosepipe watering is satisfactory for both border or container growing, provided the pressure is reduced to the minimum compatible with sufficient outlet.

Watering cans and internal reservoirs. The gardener who grows only a few plants frequently relies on a can, replenishing it from a tank or container that is kept filled. There is some virtue in this method, as it allows bulk mixing of liquid fertilisers, but obviously can watering is laborious work and it is much more suitable for container growing systems than for border culture. Hygiene is an important issue when tanks and tubs are used to store water, as these can frequently become contaminated with algae and disease. Apart from this the dilution rate tends to vary as the tanks are seldom completely emptied before being replenished.

Dilution of Liquid Fertilisers

The use of fertilisers in liquid form is almost essential for container growing systems, and while the small-scale bulk mixing referred to above may be acceptable for small-scale culture, it would be completely unsuitable for large-scale growing. Dilutors are devices into which the 'stock', or concentrated solution of the liquid fertiliser, is put. Adjustment is usually possible so that a specific dilution can be achieved, although in practice some diluters, due to various factors such as differing water pressures, tend to be erratic in behaviour or cannot be linked to mains supplies. All liquid fertilisers must be accurately measured and diluted, otherwise plants can be damaged.

Carbon Dioxide Enrichment (for Tomatoes)

The technique of artificially enriching the atmosphere with additional carbon dioxide (CO_2) has received considerable publicity in recent years, especially in the realms of early cropping. It has been shown in many cases to have a highly beneficial effect on size and quality of lower fruit trusses by increasing the speed of photosynthesis. The natural complement of CO_2 in the atmosphere is 300 parts per million (300ppm) and it is usual to give three-fold enrichment, ie to 900ppm, during daylight hours, from half an hour after sunrise till one hour before sunset. Carbon dioxide can be provided (1) by burning paraffin, though this can give rise to sulphur fumes; the use of flueless oil stoves for heating purposes does in fact provide considerable CO_2; (2) by the burning of propane or natural gas in special burners; (3) by the use of 'dry ice'; (4) by the use of liquid CO_2 dispersed from a tank through perforated polythene tubes.

The requirement for CO_2 is in the region of 100–130lb per 1,000sq ft for the propagation period, and 700–800lb per 1,000sq ft for the growing period. Obviously there will be a varying requirement according to the level of ventilation demanded by solar radiation, as CO_2 enrichment will not proceed simultaneously with ventilation.

One of the problems with CO_2 enrichment is the correct measurement of its level, and this demands specialised checking equipment. Owing to the degree of precision necessary, it is unlikely that many amateur gardeners will purposely set out to provide extra CO_2, although those wishing to do so can certainly obtain the necessary information from fuel-supplying bodies and advisory services. The culture of tomatoes under straw bale systems results in considerable CO_2 supplementation from the decomposing straw.

ADDITIONAL EQUIPMENT

Benches and Bench Warming

Benches are a very necessary part of tomato culture when plants are self-raised. Open benches made of superior pressure-treated or painted wood slats, $2\frac{1}{2}$, 3 or 4in × $\frac{1}{2}$in, with $\frac{1}{2}$in spacing, are built at a height of 30in and are ideal for tomato raising, especially if there is a pipe heating system some 9–12in below the bench. Heat can readily pass through the bench and between the plants.

Solid benches constructed of asbestos or corrugated iron, with a layer of ashes, peat or sand are however frequently used in the interests of reducing watering, and provided that there is a space between the bench and the outside of the greenhouse to allow the passage of heat, they cannot be seriously faulted. Angle steel components are excellent for bench construction especially if used with expanded metal or mesh sheets.

There is much to be said for the provision of soil-warming cables, either mains voltage or low voltage, set in 4 or 5in layer of sand for tomato propagation on a smaller scale. The use of a purpose-made or modified propagating case, or even a polythene tent, over a soil-warmed bench would commend itself to the small-scale grower wishing to provide the necessary propagating temperature for tomatoes at a lower cost, as the air temperature of the greenhouse can then be considerably reduced.

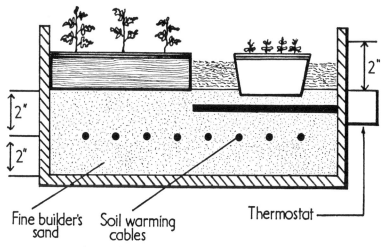

Fine builder's sand   Soil warming cables   Thermostat

11   *Make-up of an electrically warmed bench, suitable not only for tomato propagation but for other purposes*

Specifications for bench warming. Solid bench of corrugated iron or asbestos with 6in sides. Fine sand laid to 2in depth and mains voltage cable laid lengthwise at 2–4in apart to give loading of 8–12 watts per square foot. A 2in layer of sand is placed over the cable. Control by an 18in rod or phial and capillary-tube thermostat, across the run of the wire just below the surface of the bench. Use peat between the boxes or pots to cover the bench and prevent needless heat loss.

Growing Rooms

Tomato plants are frequently raised commercially in specially constructed growing rooms where high light levels are provided, temperature being controlled by fans and heaters. The heat from the lights in most instances is sufficient to maintain the temperature of the room, except during very cold weather. Seedlings, after germination in a germinating cabinet or greenhouse, are pricked off into small pots and then placed in the growing room for light treatment.

Specification for growing rooms. This is a specialised matter best taken up with the appropriate electricity board (see *Electric Growing* published by the Electricity Council; see also Chapter 6). Temperatures are 78° F (26° C) 'day' (12–16 hours) and 66° F (19° C) 'night', for 14–21 days according to light levels, which should ideally be 1,000 lumens.

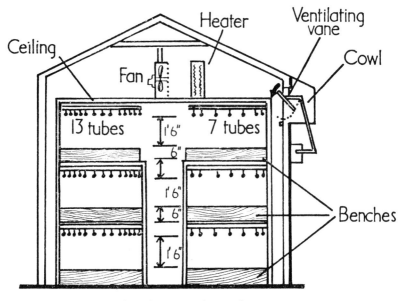

12   *Section through an original type of growing room. Linear designs of growing rooms are now used commercially.*
(*Electricity Council*)

Lighting

Apart from a working light provided by tungsten filament lamps, which would be essential for the enthusiastic grower, there is increasing interest in the use of supplementary lighting for tomato growing, especially for early cropping, in order to speed up photosynthesis in the young tomato plant and particularly during poor light periods. This technique is by no means new but has been the subject of some controversy for a great many years and recommendations have recently been somewhat modified.

Specification for supplementary lighting. HLRG mercury fluorescent reflector lamps 2·5ft above bench, 4ft apart. Alternatively MBFR/U lamps suspended 3ft above bench at 4ft apart can be used. Both types of lamp can either be permanently fixed or on a sliding rail to allow batch treatment. Either type will cover 200 seedlings. Fluorescent tubes can also be used, and are now proved as efficient. Lighting period should not be in excess of 17 hours in each 24, the plants having 7 hours' darkness, over a period of 17–21 days (see also Chapter 6). Greenhouse temperature should be 68° F (20° C) by day and 60° F (16° C) at night.

Capillary Benches

These provide for a constant supply of water ½in below the surface of coarse sand, which is spread out in a 2–3in layer in a perfectly flat polythene-lined 'basin', the water level being controlled by a ball-cock tank set alongside the bench (fig 13) or a float controller. Such benches are used mostly for growing rooms in tomato culture.

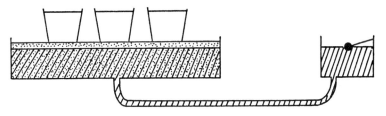

13   *Capillary benches can readily be made up by using polythene-lined basins and any suitable method of water-level control. They have a limited use in tomato propagation*

Lining for Greenhouses

The lining of greenhouses with light gauge polythene is a useful technique for reducing heat loss. Polythene lining cannot be compared with double glazing, where two hermetically sealed panes of glass provide a 'dead' layer of air between them. It does, however, considerably reduce the heat loss through overlaps in the glass or loose glazing systems and at the same time prevents draughts. The quoted reduction in heat loss is from 1·1 to 0·62Btu per square foot.

Lining with polythene is usually carried out only partially, vents being left unobstructed, and the net effect is a slightly higher average temperature in the greenhouse for a lower artificial heat input. A raising of the humidity occurs due to moisture condensation on the polythene and this, while advantageous for propagation, can be slightly disadvantageous for full season growing, depending on the efficiency of the ventilation. For this reason polythene lining is perhaps best employed only on a limited scale for tomato growing, preferably along the side of a greenhouse subjected to cold winds or in the region of doors.

Specification for lining. Light gauge polythene is tacked on with drawing-pins or staples through paper pads, to astragals or wooden strips fitted into the space that frequently exists between the extruded alloy glazing bars. Vents and doors can be fitted with movable curtains of polythene or left unobstructed.

4

Growth and Nutritional Needs

PHYSIOLOGICAL ASPECTS OF GROWTH

All green plants depend on supplies of air, moisture and nutrients, and thereafter, if there are no inhibiting factors, growth proceeds at a temperature and light level suited to the metabolism of the plant in question. It has been agreed that the particular metabolism of the tomato plant demands constant warmth and a high light intensity, and the gardener or grower must therefore provide these conditions; at the same time attention must be paid to the tomato's gross appetite for essential elements.

The complexities surrounding growth are discussed fully in any works on botany or plant physiology, but it will perhaps help us to understand the physiology of tomatoes better if some of the more important issues are briefly referred to.

Respiration

The function of breathing is carried on by all living organisms. Air is taken in through the pores of green plant leaves, and to a lesser extent the stems, the oxygen content in the air is extracted as energy for various chemical processes and the carbon dioxide is expelled. The rate at which respiration takes place depends on the age of the plant—a young plant breathes faster than an old plant—the temperature, and of course the rate of growth. Respiration is, however, catobolic (destructive) and continuous, an important thing to remember in tomato growing, in respect of day and night temperatures and the using up at night of carbohydrate manufactured during the day.

Process of respiration. Carried on continuously:

47

oxygen extracted for
Air———→ energy and various ———→carbon dioxide expelled
chemical processes

Photosynthesis

During photosynthesis air is taken in by the plant through its pores, the carbon dioxide contained in it at 300 parts per million (or more) is extracted and the oxygen is expelled—the direct antithesis of respiration. Photosynthesis takes place *only* in the presence of light and with the aid of the catalyst chlorophyll, the green pigment present in all green plants. As photosynthesis is responsible for the formation of the carbohydrates essential to sustain growth, there is a direct relationship between the rate of photosynthesis and the rate of growth, including the swelling of fruit. The speed at which photosynthesis takes place in the presence of adequate water, carbon dioxide and nutrients depends largely on temperature and light level. Modern research suggests, however, that there are times when there is insufficient carbon dioxide present in the air to allow photosynthesis to take place rapidly enough to cope with the requirements of the plant. The research work carried out with carbon dioxide has given some spectacular results in many cases, resulting in larger and earlier yields of high quality fruit, although there is still some conflict of views on the economics of its use.
Process of photosynthesis. Proceeds only in the presence of light and its rate is governed by light and temperature levels.

Air———→carbon dioxide extracted———→oxygen expelled

Transpiration

Water absorbed by the roots (see osmosis, below) rises up through the stem, being passed from cell to cell through tissue called the xylem, and travels into the leaves. Water that is superfluous to the needs of the plant is expelled as water vapour through the pores in the stem and leaves, the evaporation process allowing the plant to keep 'cool', simultaneously a transpiration stream is created capable of transporting dissolved nutrients through the plant tissue.

It is essential for the tomato grower to appreciate that the rate of

Page 49 (Above): The author's Dutch light greenhouse with a crop of tomatoes. Those on the left are on straw bales; those on the right in border
culture. (Below): A wide-span glasshouse specially designed for maximum
light catchment and sited east-west, in the course of erection at Eaglesham,
Renfrewshire

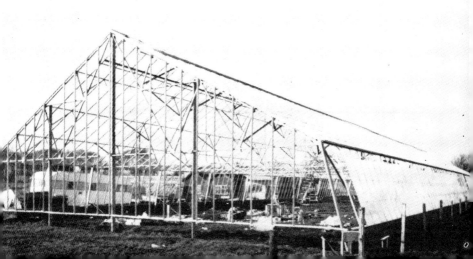

Page 50 (*Above*): Bubble greenhouses, while excellent for most crops, pose problems of support for tomatoes. (*Below*): An earlier design of commercial plastic greenhouse; the do-it-yourself enthusiast could easily produce something similar on a much smaller scale. Note the plastic windbreak in the foreground

water uptake (and nutrients in solution) is dependent on (1) the healthy state of the roots, (2) the health and condition of the xylem tissue of the plant, (3) the total leaf area of the plant. When transpiration rate is excessive in a hot dry atmosphere, and especially when there is a large amount of foliage, the plant cells partially collapse, normally recovering at night when it is cooler and the rate of transpiration slows up. This is especially the case when the temperature suddenly rises, following a relatively cool period of several days. The rate of uptake also depends (4) on the temperature of the air, (5) the humidity of the atmosphere (the rate of water vapour discharge is directly related to the humidity of the air and the respective pressures inside and outside the plant) and (5), perhaps most important of all, on the respective osmotic pressures in the soil and the plant (see below).

Process of transpiration. Water is taken into the plant through the process of osmosis, passed up to the leaves, and a proportion is given off as water vapour at a rate dependent largely on temperature, humidity, and the respective osmotic pressures of plant and soil.

Osmosis

This important process is one which must be fully understood by tomato growers, particularly those concerned with the more sophisticated forms of culture now practised (these are dealt with in their respective chapters).

Osmosis is a process whereby a high concentration of salts in solution on one side of a semi-permeable membrane absorbs the *water* of a less concentrated solution through the semi-permeable membrane. In simpler terms, the stronger concentration in the cell of the root of the plant 'pulls' on the less concentrated solution contained in the soil. Note that, technically speaking, osmosis only pulls in the *water*, it being stated by plant physiologists that the solution of nutrients enters the plant by diffusion, a separate process though closely associated with osmosis. The movement of the solution from cell to cell in the plant goes on by an osmotic chain reaction, the cell which has received water having its salt concentration lowered, thus allowing the now more concentrated solution in the next cell to pull in its solution, and so on. The rate at which the uptake of water and nutrients takes place, while partly dependent on the rate of transpiration, also hinges largely on the ability of the plant to take in water initially by osmosis. Obviously if there is near equilibrium because the salt solution of the soil is altered by the application of concentrated fertilisers (liquid or solid), the plant is unable to absorb its solution freely. Indeed, if excess fertiliser is applied to the soil, reverse osmosis may take place, causing serious damage to plant tissue, this being the way in which some weed-killers act. By exercising strict control over the dilution of liquid fertilisers applied to the soil or growing media, it is possible to exert a marked influence on the growth rate of the plant. This is of considerable importance for the culture of early tomatoes, which often tend to make vegetative instead of productive growth.

Process of osmosis. The stronger salt solution of plant cell, ie in the root hair, pulls in the weaker solution from the soil and there is a chain reaction process throughout the plant. Control of soil solution concentration can exercise in turn a considerable control in the growth rate of the plant.

Translocation

This is the movement of the salts and manufactured foodstuffs around the plant, largely through tissue known as phloem. Foods manufactured in the leaves are obviously required for various functions throughout the plant tissue, particularly for the swelling of the fruit. Irregularities in watering, temperature or ventilation can disrupt the transport of foodstuffs, resulting in various troubles such as uneven fruit ripening, 'black bottoms' on fruit, leaf distortions and many other symptoms. Unbalanced nutritional conditions in the soil are usually highly contributory to the various irregularities.

Pollination and Fertilisation

The sexual reproduction process carried out by all flowering plants depends on the male pollen grain reaching the stigma of the female (pollination), the subsequent germination of the pollen grain and the growth of a pollen tube so that fusion between the male gamete and the female ovule takes place, resulting in *fertilisation*. The flowers of the

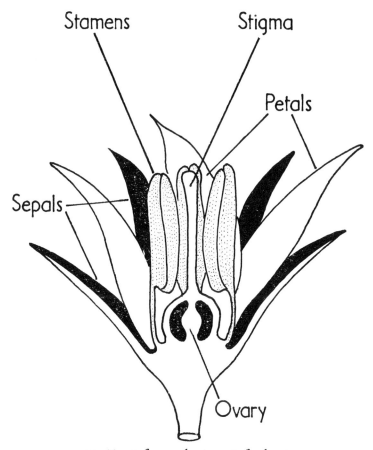

14   *Tomato flower—showing main floral parts*

tomato plant, which contain both male and female organs, are so shaped that self-pollination readily occurs (fig 14) although some assistance is welcomed. Tapping with a cane or shaking by applying droplets of water helps considerably. A reasonably humid atmosphere will also ensure that the pollen grain does not dry out before it germinates, this being the main reason why damping-down is practised so frequently. Partial or complete failure of fertilisation is due to various causes, including infertile pollen, and will result in small 'chat' fruits or the complete dropping of the flower by shrivelling of the stalk, which lacks the necessary stimulus when fertilisation is unsuccessful. Fertile

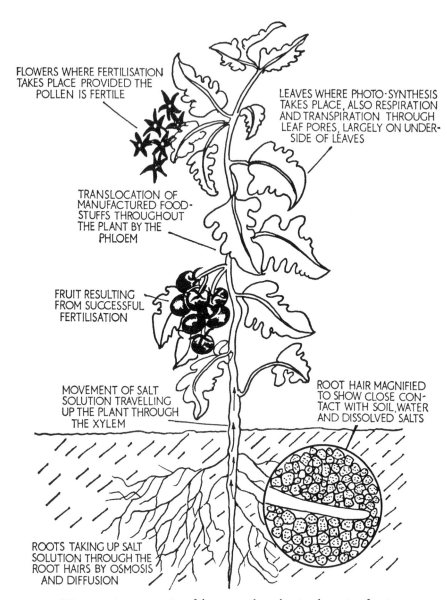

FLOWERS WHERE FERTILISATION
TAKES PLACE PROVIDED THE
POLLEN IS FERTILE

LEAVES WHERE PHOTO·SYNTHESIS
TAKES PLACE, ALSO RESPIRATION
AND TRANSPIRATION THROUGH
LEAF PORES, LARGELY ON UNDER-
SIDE OF LEAVES

TRANSLOCATION OF
MANUFACTURED FOOD-
STUFFS THROUGHOUT
THE PLANT BY THE
PHLOEM

FRUIT RESULTING
FROM SUCCESSFUL
FERTILISATION

MOVEMENT OF SALT
SOLUTION TRAVELLING
UP THE PLANT THROUGH
THE XYLEM

ROOT HAIR MAGNIFIED
TO SHOW CLOSE CON-
TACT WITH SOIL, WATER
AND DISSOLVED SALTS

ROOTS TAKING UP SALT
SOLUTION THROUGH THE
ROOT HAIRS BY OSMOSIS
AND DIFFUSION

15   *Diagrammatic representation of the tomato plant, showing the various functions*

pollen is only produced when there is a sufficiently high light intensity during truss initiation, and when temperature and level of nutrients have been satisfactory. These matters will be referred to in more detail in Chapter 6 and elsewhere.

PLANT NUTRIENTS AND THE TOMATO

Plant nutrition is a very complicated and involved subject, and to discuss it in real detail is beyond the scope of this book. We have alluded briefly to the methods by which the green growing plant obtains its nutrients—by osmosis, gaseous exchange and diffusi n—and these processes are basic to all plant growth. Some elementary understanding of plant nutrition is, however, essential if one is to exercise any control over the rate and precise manner in which the plant grows and to coax it to produce the maximum amount of fruit—and this is particularly so with tomatoes.

What are Plant Nutrients?

Essential elements are required by all living organisms to provide the energy needed for all the various processes concerned with growth. We devour our food, but the plant is not able to digest solid elements; it can only take them when they are either dissolved in water, which itself contains essential elements (oxygen and hydrogen), or in the form of a gas, in the case of carbon and oxygen.

Once dissolved, these elements can be taken in either by the roots of the plant or absorbed through the leaves, provided the chemicals are not of a form which would cause physical damage to the leaves.

The Source of Plant Nutrients

All soils are derived from the long-term weathering of the rocks which form the crust of the earth. Over the course of a great many years what is known as the soil has developed on the surface of the earth. It consists of a heterogeneous collection of broken-down rock, the particle size of which varies from gravel through sand, silt and clay. Intermingled with the mineral particles are plant and animal remains and a vast population of micro-organisms and larger forms of life such as worms, all with differing roles.

The precise nature of a soil is classified largely by the particle size: 'light' sandy soils are composed predominantly of larger particles, 'heavy' clay soils of small particles, and 'medium' soils of a mixture of both. Beneficial bacteria generally thrive in well-aerated soils and carry on their various roles of rendering nutrients palatable to plants. In the case of proteins these are reduced to nitrogen—coupled with which there is intense chemical activity to render all plant nutrients soluble, largely through the acid formed by bacterial and chemical action. Much of this takes place on the large surface area of the mineral particles, and it is desirable to build up crumb structure where the fine particles adhere together, as a result of the formation of organic gums, which arise from the activities of micro-organisms; electrolytic action also brings sub-particles together.

The component constituents of rocks vary considerably throughout the world, but to our knowledge invariably include silicon, aluminium, phosphorus, potassium, calcium, magnesium, sulphur, boron, manganese, copper, zinc, iron, molybdenum, chlorine, sodium, iodine, cobalt, selenium. Other elements essential for plant growth are hydrogen and oxygen, which are 90–95 per cent supplied by water, and nitrogen derived from the atmosphere by fixation and precipitation, and of course by the breakdown of organic matter. Carbon is also essential, this largely being taken up in gaseous form from the atmosphere.

Where soils are used for the culture of tomatoes, there will already be a limited natural source of elements and plant nutrients available to the plants. Where soil-less mediums are used, nearly all the nutrients must artificially be supplied in palatable form by the addition of readily available fertilisers, and this is a matter which will be discussed further in Chapter 5.

The Function of Plant Nutrients

When discussing plant nutrition and the precise role of each element in the general metabolism of the plant, it is easy to fall into the trap of thinking, in compartmentalised fashion, that one element is responsible for only one function. Such is certainly not the case, as the whole process of growth is a most complicated and interrelated chemical process involving integration over the whole range of elements which are necessary for successful growth, but obviously in quantities which vary between different species of plants.

The only practical way of discussing plant nutrition is, however, to give more or less individual consideration to the elements concerned, and this we shall now do with particular reference to tomato culture (see also Chapter 12). But it must be remembered that each and every element is a cog in the wheel of growth and that if one is missing a breakdown in growth can occur.

Nitrogen (N)

Nitrogen is present in all forms of plant material capable of life. The sphere of growth where nitrogen appears to play the largest part is by overall increase of total bulk. With tomato plants an excess of nitrogen can be an embarrassment by inducing soft and relatively unproductive growth, this being particularly true when light intensity is low, which can often be the case early in the year. The lush green colouration and excessively curled foliage caused by an excess of nitrogen show quite clearly that it is much concerned with the process of photosynthesis carried on in the leaves.

Nitrogen is taken up by the plant as highly soluble nitrate, usually having become converted by soil bacteria from ammonia to nitrite and then nitrate. Plants can also absorb liquid ammonium in limited quantities. Where tomatoes are grown in soil-less media it is necessary to provide a form of nitrogen which is immediately available, without recourse to bacterial action initially.

The availability of nitrogen to the plant in soil-containing media will obviously be largely dependent on the rate of bacterial activity, and this will depend on soil temperature and also on whether the bacteria are inhibited or otherwise by organisms such as engulfing protozoa, by waterlogging, lack of air, consolidation or other factors. The flush of nitrogen which will invariably result following partial sterilisation of soil (see Chapter 13) is due to the ability of the thick-walled ammonifying bacteria to survive normal sterilisation temperatures in contrast to other types, including bacteria-engulfing protozoa, which are killed off. Pest and disease control is of course the main reason for sterilisation, the dissipation of plant toxins also being an important issue.

The constant breakdown of the organic matter, which goes on continuously, does of course result in the formation of carbon dioxide (CO_2) which, with water, forms a weak acid solution in the soil, and it is this weak acid which is one of the main agents in dissolving not only nitrogen, but the other elements. The highly soluble nature of nitrogen makes it extremely vulnerable to leaching (washing out), which is one reason why watering should not be carried out to excess in glasshouse culture.

Excess of nitrogen in tomato plants is usually shown quite clearly by lush unproductive growth, and shortage by pale, often yellow, small-leaved lank plants. Plants taking up too much ammonium show varied symptoms, 'black bottoms' on fruit is thought to be one of these (see page 62). Nitrogen and ammonia are very soluble, hence the possible need for flooding after heat sterilisation (see Chapter 13).

Summary. Nitrogen is needed in large amounts by tomatoes.

Excess: lush soft growth.

Shortage: pale small leaves, often yellowed.

Phosphorus (P)

Phosphorus is associated with all life and is a basic constituent of every living cell, whether animal or plant. Plants generally require considerable amounts of phosphorus, and yet the tomato is not particularly demanding in this respect, especially if we look at the quantities of potassium and nitrogen needed. Natural supplies of phosphorus exist in many types of soil, but there can also be a deficiency of phosphorus, not necessarily because of actual shortage, but because what there is is in an unavailable form.

The net effects of phosphorus on the tomato plant are so all-embracing that it is difficult to single out its precise function. As phosphate is essential for the whole process of life, it is highly essential for all root growth and tomatoes are no exception. Young plants benefit particularly from it, but the benefits persist right to maturity by hastening flower and fruit development.

Phosphorus is available to plants as phosphoric acid when it has been dissolved in the weak acid produced in the soil by various processes. The problem is that phosphorus is highly insoluble and, of the total present, only a very small proportion is actually available to plants. In acid soils much of the phosphate combines with iron and aluminium to form insoluble compounds, which shows the need to keep soils well supplied with calcium (lime), although liming is not the full answer as insoluble calcium phosphate can be formed. It is thought that when phosphorus is added to a soil the actual phosphorus applied becomes insoluble but releases a quantity of previously unavailable phosphorus for use. Observation of the quantities of phosphate found by analysis to be present in older tomato soils in considerable quantities, shows quite clearly that tomato plants themselves are able to extract considerable amounts of phosphorus from the soil reserves—so much so that it may be unnecesary to apply any phosphorus at all for many years when cropping such soils.

Lack of phosphorus in the tomato plant is indicated by a bluish colouration, this symptom frequently being exhibited by chilled plants unable to assimilate the phosphorus they properly require. Symptoms of excess phosphorus are difficult to define, possibly because amounts surplus to plant requirements are either lost by drainage or alternatively become unavailable.

Summary. Phosphorus is only needed in limited amounts by tomatoes.

Shortage: bluish colouration, lateness of general development.

Excess: difficult to detect visually.

Potassium (K)

Like the other main elements, potassium is an essential constituent of all living matter, being required in very large amounts, especially by the tomato. Potassium is a constituent of many types of rock and is generally present naturally in larger quantities than nitrogen and phosphorus, yet only a small proportion may be available to the growing plant. It is taken up in solution as potassium oxide (K_2O), when it seems to be largely connected with carbohydrate formation and the rate at which photosynthesis occurs.

MAIN TYPES OF BULKY ORGANICS USED
IN TOMATO GROWING

(Most of these organic materials contain trace elements in addition
to the main elements stated)

Average quality	% in fresh samples			Comments
	Nitrogen (N)	Phosphorus (P₂O₅)	Potash (K₂O)	
Farmyard manure	0·43	0·19	0·44	Farmyard manure is found to vary considerably and for tomato culture should be well decomposed. Contains reasonably balanced quantities of the main nutrients.
Poultry manure	2·1	1·21	0·60	Note the high nitrogen content (richer than farmyard manure) which gives rise to problems when used too liberally for tomatoes.
Sewage sludge	2·32	1·29	0·25	Useful for tomatoes grown out of doors. Nitrogen content not in such an available form as in farmyard manure.
Seaweed	0·4–0·8	0·1–0·2	1–2	A useful form of organic matter fairly rich in potash. Should be used fresh or composted.
Peat	0·7–3	0·1–0·2	0·1–0·3	Peat is used mainly to 'condition' soils intended for tomato cultivation, also as top dressing or mulch, and of course as the main or sole ingredient of soil-less composts (see page 83). The pH of sphagnum peat is about 3·5–4·0. The nutrient content of peat is released only very slowly.

Potassium is very important for flowers and fruit, and therefore very necessary for tomatoes, which form a high proportion of fruit in relation to total plant bulk. Potassium improves quality and flavour and imparts a measure of disease resistance to plants, in this way contrasting with the effects of nitrogen. This means that a shortage of potassium invariably results in lush soft growth with almost a bluish hue and with marginal leaf necrosis. Research has shown that potash has a great effect on the regulation of water uptake. Much of the success in tomato growing lies, in fact, in supplying the correct balance of nitrogen and potash at the right time.

Summary. Potash is needed in large amounts by the tomato.

Shortage: lush soft growth, marginal leaf necrosis.

Excess: hard stunted growth, very dark green colouration.

Magnesium

This is a highly essential element for tomato plants, being closely tied up with chlorophyll formation, and when it is lacking or unavailable the photosynthetic process is affected, with the result that browny-orange areas appear on the older leaves between the veins, especially in the sunniest parts of the greenhouse. Magnesium also acts as a carrier for phosphate and regulates the uptake of other nutrients, including the translocation of carbohydrates around the plant. It can be seen, therefore, that magnesium has a major role to play in tomato culture, and for this reason it cannot generally be taken for granted that there is sufficient magnesium present in the soil.

Summary. Magnesium is required in fairly large quantities by tomatoes.

Shortage: interveinal chlorosis (yellowing) of leaves occurs on lower leaves first, and gradually travels up the plant. Orange areas usually develop later. The sunny side of the greenhouse is worst affected. Plants grown in light types of soil where potash has to be used frequently generally suffer most from magnesium deficiency.

Excess: difficult to define, although surplus magnesium generally affects salt content of soil.

Calcium

Calcium is present naturally in various forms, particularly in phosphate-containing soils. Apart from neutralising acid excretions of plants and stimulating bacterial action, calcium is a constituent of all plants, especially of the cell walls. Calcium has an important part to play in the base exchange mechanism which operates to supply plants with nutrients. It also serves to bring together the fine particles of clay to form crumbs, a process called flocculation. The amount of exchangeable calcium in the soil is measured by what is called the pH scale, which ranges from 1 to 14, 7 being the neutral point between sourness (low pH figures) and sweetness (high pH figures). (See page 70.)

For tomato growing in soil-containing media, the pH figure should be between 6 and 6·5, but for growing in soil-less media where there is less bacterial action, at least initially, lower pH figures seem to be quite acceptable. The amount of exchangeable calcium in the soil can greatly affect the availability of other elements, and vice versa, especially ammonia and potash, which can shut off or immobilise the calcium (see page 197). Too low a pH figure can result in toxicity of manganese, too high a figure in unavailability of manganese.

Summary. Calcium is required in reasonable quantities to tomato plants.

Shortage: general lack of vigour, usually with yellowing symptoms typical of nitrogen shortage or whitening at growing points.

Excess: difficult to define, although iron shortage is main symptom of a high pH figure (see below).

Iron

Iron is much concerned with the photosynthetic process and shortage of it usually turns leaves pale, and in severe cases white. Iron is relatively insoluble at normal pH figures, and becomes more so at high pH figures, when insoluble salts may be formed. Excess iron is supposedly toxic to plants, although I have personal experience of considerable excess iron being applied in error to tomato plants with no ill effects! But excess application is not to be generally recommended.

Summary. Iron is an element essential to tomatoes in small amounts.

Shortage: yellowing or whitening of leaf.
Excess: difficult to define accurately.

Manganese

Again, this is much concerned with photosynthesis, and shortage of it causes chlorosis or mottling of the leaves. Manganese deficiency can often occur at high pH levels with young rapidly growing plants, and can disappear once the plant's growth rate decreases. Manganese toxicity is far more important than deficiency, causing an intense blue-black colouration at growing points and drooping down of leaves. It can occur in acid soils.

Summary. Manganese is an essential element in tomato culture—small amounts only required.

Shortage: chlorosis and mottling of leaves, especially when young.

Excess: blackish colouration of growing points, coupled with a drooping appearance.

Boron

Although at one time boron was not thought to be an important element in tomato nutrition, it has attracted considerable attention in recent years and must now be considered as having a vital role to play in the general growth of plants, deficiency causing shrivelling and browning of growing points. In some cases a brown corky layer can form below the skin around the fruit. Usually sufficient quantities are available in the soil for all normal needs.

Summary. Boron is required in small amounts, but its presence is essential.

Shortage: shrivelling of growing points and corkiness under the skin of fruit.

Excess: unlikely to occur, but would cause marginal shrivelling on leaves.

Sulphur

Sulphur has an important part to play in general plant nutrition, but it is difficult to be precise about its exact function. As sulphur is present in many fertilising materials, its shortage is seldom a problem.

The other elements referred to on page 56 are unlikely to give rise to any severe complications with tomato plants under normal cultural conditions (see spectrochemical analysis, page 71).

TOMATO BASE FERTILISER FORMULATIONS

Proprietary base fertilisers are available in standard, high potash and high nitrogen forms. Details of composts and soil-less media are discussed in Chapter 5.

An *average* analysis is as follows:

Base fertiliser	Nitrogen % N	Phosphorus % P_2O_5 Sol	Insol	Potash % K_2O
Standard	9·5	9	0·5	13·2
High potash	6	10	0·5	17·5
High nitrogen (seldom necessary)	12	5·5	0·5	6
John Innes base (see page 84) use as high potash	5·2	7	0·5	10

The use of base fertiliser is discussed in Chapter 8, but generally speaking high potash base is used following the sterilisation by heat of soil-containing media, or for vigorous varieties on unsterilised soils. Standard base is used when the growing media is fresh or has been chemically sterilised. High nitrogen base is used for soils thought to be deficient in nitrogen.

FERTILISER USE IN TOMATO CULTURE

Type	N	P_2O_5	K_2O	Other elements	Comments and rates of use
Hoof and horn meal	12–14%				An organic form of nitrogen much used in tomato base fertiliser and composts. Release of ammonia can be rapid, but complete release of nitrogen content is over a period. Use up to 4oz per sq yd.
Dried blood	7–12%	1–2%	1%		Main use is as 'safe' top dressing for tomatoes at 1–2oz per sq yd as a quick source of nitrogen. Although organic its release of nitrogen is rapid. Also useful diluted at up to 2oz per gal.
Nitro-chalk	25%			Carbonate of lime	Inorganic quick-acting sources of nitrogen. Main use for tomatoes requiring an urgent 'kick'. Frequently used where virus 'check' occurs (see Chapter 12). Use at rates of ½–1oz per sq yd (Percentages subject to change.)
Nitra-shell	34%				
Nitram	34.5%				
Nitro 26	26%				
Nitrate of potash or potassium nitrate	12–15%		42%		Inorganic convenient quick source of nitrogen. Best used as a liquid fertiliser (see page 163).
Ammonium nitrate	35%				Inorganic. Used as a constituent of liquid fertilisers (see page 163).
Urea	46%				Organic. Used mainly as a constituent of liquid feed (see page 163). It can be used 'dry' at 1oz per sq yd as organic source of nitrogen. The more sophisticated form is urea formaldehyde (Nitroform) 38% N, releasing the nitrogen over a period. Use as directed.
Ammonium phosphate	11%	48%			Inorganic. Used mainly in slow-release-type fertilisers, seldom as a separate entity.
Bone meal	1–5%	15–32%			Organic. Not used greatly in tomato culture but would be useful in a soil very low in phosphates, applied in addition

Type	N	P_2O_5	K_2O	Other elements	Comments and rates of use
Super-phosphates of lime		18–21%			to base feed. Use at 3–4oz per sq yd. Inorganic. Best form of phosphates for tomato culture, used generally as a constituent of base feeds and in composts. Can be used at up to 4oz per sq yd.
Sulphate of potash			48%		Inorganic. Use at 1–2oz per sq yd as quick source of potash. Can also be diluted in water at 1oz per gal.
Epsom salt				Magnesium oxide 16%	Inorganic. Use up to 3–6oz per sq yd for tomatoes. Used also as foliar feed at 2lb per 10gal with a spreader.
Lime (hydrated) (ground) (magnesian)				Calcium oxide 68% 47% Magnesium oxide 10–40% Calcium oxide 50–80%	The main form of lime used for tomato culture is *ground limestone*, a relatively slow acting form of calcium. *Magnesian limestone* is frequently used for tomato growing as it contains magnesium. Both forms are used according to soil analysis but generally at 8oz per sq yd. For quantities used in soil-less media see Chapter 5.
Calcium nitrate	15%				Useful source of nitrogen *and* calcium.

Page 67 (*Above left*):
A free discharge warm-
air heater (oil-fired) for
use in large green-
houses. (*Above right*):
Fan ventilation has
many advantages for
positive air change in
greenhouses. The fans
must be louvred to
prevent free air entry of
air when they are not
in operation.
(*Left*): An aspirated
screen housing a
thermostat is an essen-
tial item for accurate
temperature control

Page 68 (*Above left*): Spraylines can be used at high or low levels for water and liquid feed application—in the latter case when the lower fruit has been picked. (*Above right*): Trickle irrigation systems are useful for border or container growing. (*Below*): Plants set out in 9in whalehide pots on a deep layer of ashes during the initial stages of ring culture

5

Nutrients and Growing Media: Specifications

UNDERSTANDING SOIL AND TISSUE ANALYSIS

Soil analysis in its various forms comes in for much criticism by soil chemists, who consider that most systems of analysis do not truly reflect the quantities of nutrients actually available to the plants. Recent years have seen the introduction of different systems of reporting soil analysis figures, and also the increased use of spectrochemical analysis of both soil and tissue, and the 'spot' analysis of plant tissue.

The full technical implications of soil analysis are not likely to be grasped by either gardener or grower, and it will suffice for practical purposes to describe the various analysis systems fairly broadly.

Soil Testing Kits

These take various forms and can establish (1) the acidity or alkalinity of the soil by testing with litmus paper, (2) an approximate pH figure by means of a series of litmus-paper tests, and (3) a remarkably accurate pH figure by using indicator fluid, either separately or in combination with barium sulphate as a precipitant. Complete soil testing kits using indicator fluids are also available for pH figures and relative availability or percentage deficiency of the three main elements—nitrogen, phosphorus and potash. While they cannot be considered nearly as accurate as laboratory analysis, soil testing kits have the virtue of giving quick 'on the spot' guidance, thus eliminating the unavoidable delays in laboratory analysis.

Laboratory Analysis

This is undertaken in most parts of Britain, through official advisory bodies, societies, or commercial firms. The range and method of analyses and system of reporting vary, but the information which is normally available is set out in the following paragraphs.

Organic matter. The level of organic matter is a very useful guide to the tomato grower and is reported either as a percentage figure or simply as 'low', 'medium' or 'high'. A figure of 8–12 per cent (including moisture) would be described as a 'good' organic matter content for a tomato soil.

Lime requirement. As mentioned on page 62, the lime requirement is of considerable interest to the tomato grower, and this is given in order to adjust to a pH figure, the norm being between pH 5·5 and pH 6·5 (note that lower pH figures are acceptable for soil-less media), lime being added generally in the form of ground limestone to adjust the pH figure. As lime requirements are given per square area of soil to a depth of 9–10in, some conversion to bulk is desirable. Despite metrication the bushel (22 × 10 × 10 inches) will probably continue to be used for some time in horticulture, and there are approximately 5 bushels in a square yard (to 9in depth), 5½ in a square metre, and 22 (21·7) in a cubic yard (see p 85).

A soft fine-particled calcium carbonate or chalk is frequently used for growing tomatoes, and this has the same value as ground limestone. Hydrated lime is 'hotter' than ground limestone and only three-quarters of the amount is required for the same liming value, but it is not generally advised as it can scorch foliage. There is a time lag between the application of the ground limestone and the attainment of the pH (see page 86).

Available nitrogen (N). This is not a test carried out by every laboratory, because the nitrogen level varies greatly according to the rate of bacterial action as dictated by season, temperature, and whether soil has been heat sterilised or not. The nitrogen availability can be assessed in various ways, as ammonia, total nitrogen, etc, expressed in ppm.

Index figures are also being used by some bodies, and it is necessary to refer to tables for their significance in relation to other methods of reporting. The usual terms relating to availability—low, medium and high—are used in addition in most cases, yet correct interpretation of the results by an experienced chemist or horticulturist is essential. Tomatoes require a 'high' nitrogen figure for optimum growth.

Available potassium (K_2O or K). Stated as mgm per 100gm, or ppm, or quoted as low, medium or high, this is a very useful guide to quantities of potash available in the soil. For tomatoes the figure should be 60–80mgm per 100gm K_2O or 500–700ppm K (in soil), and stated as 'high' (see also soluble salt content, below). Where index factors are quoted to standard scales the figure would be 4–5.

Available phosphorus (P or P_2O_5). The figures are reported as for potash, in respect of available phosphorus, although this is a difficult figure to reconcile with true availability to plants because of the 'fixing' of phosphorus in the soil. Old tomato borders which have been cropped over a number of years frequently show a very high level of phosphorus. Ideal figures are 30–60mgm per 100gm P_2O_5 or 41–70ppm P (in soil), or index 4–5 and stated as 'high'.

Soluble salt content (pC or Cf—conductivity factor). This is a very important figure for the tomato grower, and one to which considerable attention must be paid in view of the respective osmotic pressures in the cell sap and the soil water (see page 51); pC figures of 2·8–3·0, or Cf of 16–10 (index around 3) are desirable for tomato growing, as salt concentrations exercise a means of control on growth. Higher concentrations of salt indicated by a *lower* pC figure in the order of 2·6 (Cf 25, index 5–6) are so dangerously near osmotic equilibrium that growth of the tomato plants are very likely to be adversely affected. Roots may in fact be chemically damaged if the salt concentrations are too high. Spot checks carried out regularly are very useful for commercial growers before and after planting, and many of them feel that the purchase of a salt concentration meter is worth while (see page 147). Growing media can be checked after mixing; generally all proprietary purchased media for full season growing are already at the correct pC level, or alternatively will reach the correct level if accurately self-formulated.

Spectrochemical Analysis

This determines the levels of micro-elements such as zinc, manganese, copper and iron. Both soils and plant tissue can be examined, but in each instance spectrochemical analysis would only normally be carried out by advisory bodies following abnormalities in growth.

Tissue Analysis

Here plant tissue is analysed either by spectrochemical or other method to determine the actual nutrient level of the tissue, an extremely valuable operation provided there can be comparison between healthy tissue and that showing abnormal growth. There is an increasing use of tissue analysis for quick checks on the nutrient level in the plant.

Other Methods of Analysis

It is claimed that the observation of the growth made by various fungal and other cultures introduced under laboratory conditions to samples of soil gives a much more realistic picture of the actual availability of nutrients than the normal extraction method of soil analysis.

Methods of analysis based on the utilisation of 'labelled' radioactive elements are also used in research work. By means of these it is possible to follow the progress of elements through the plant tissue.

But the normal visual check on growth still remains a highly important procedure in growing, and this is especially true in respect of the tomato plant, which shows a quick reaction to excesses and shortages of main and micro-nutrients.

Eelworm Determination

Soil infested with the cysts of the female *Heterodera rostochiensis* (potato eelworm) which contain viable larvae should preferably not be used for tomato culture. The crippling effect of potato eelworm on tomatoes is discussed in Chapter 12; it is obviously important to know whether or not a soil contains eelworm cysts. An eelworm count giving the number of eelworm per gramme of soil and whether they contain live larvae, is usually available from advisory services.

Disease, Pest and Weed Potential of Soil

Detection by pathological means of the presence of tomato diseases before a soil is used for tomato growing is not a very satisfactory procedure, although it can be done by examining root debris of the previous crop. The best guidance is, however, the previous cropping history of the soil. Many pests can be visually detected, especially wireworm, and the same is true of the vegetative portions of bad perennial weeds.

The Importance of Representative Samples

Soils sent for analysis should be entirely representative of the main bulk of the growing medium. Where border soils are concerned, several small samples should be taken with a trowel, auger, or other instrument which will conveniently lift a small quantity of soil from a depth of 5–6in. The total bulk of this sample should weigh about 2lb, and should be placed in a polythene bag and clearly labelled for dispatch to the advisory body. One 2lb sample will generally be suitable for an area of up to 1,000sq ft, although there should be no hesitation in taking further samples of 'good' and 'bad' areas.

Growing media may be sampled throughout the mixture by lifting small portions with a trowel to bulk into a 2lb sample. Samples are best taken from the *formulated growing medium*, although there is a lot to be said for sampling doubtful ingredients *before* mixing. This applies especially to sand, loam, or very acid peat. *Always label samples clearly*.

Fertiliser and Liquid Feed Analysis

The analysis of fertilisers is generally unnecessary, as the quality control is very strict at the manufacturing stage. Analysis can be useful in the case of doubtful identity. A frequent form of analysis is to check the salt concentration (pC or Cf) of liquid fertilisers *when* diluted and ready for crop application.

THE NUTRIENT REQUIREMENTS OF TOMATOES

Estimating Nutrient Needs

Estimating quantities of nutrients or using soil analysis figures to cater for the actual nutrient need of tomatoes either in borders or in formulated compost is not exactly a simple matter, nor can it even be known whether all the nutrients given are available to the plant. A further complication is that the differing performance of each plant makes its own special demands, which in simple terms means that a tomato plant yielding only 6lb of fruit will require fewer nutrients than a plant yielding 10lb of fruit. The stated requirements for the major nutrients for different cropping levels are subject to some variation, but fall broadly into the following levels for a 50–60 ton crop per acre (which is 9–10lb per plant):

> 800–1,000lb potash
> 400–500lb nitrogen
> 75–100lb phosphorus
> 110–130lb magnesium
> 500–600lb calcium

Even allowing for some inaccuracy and a much lower crop of 5–6lb per plant, these figures show quite clearly that potash, calcium, nitrogen, magnesium and phosphorus are required in considerable quantities, each in amounts which are relatively consistent. The graphs opposite reduce these figures to the requirements for individual plants. Calcium is not included, as it is best supplied by adjustment of the pH figure; the need of soil for calcium varies considerably. Note that lower pH figures are acceptable with soil-less culture systems.

Allowing for drainage loss in use of 15–20 per cent or more, and for the 'fixing' of phosphorus, it is safe to assume that about 50 per cent of the nutrient applied is available to the plants in nitrogen and potash and 20 per cent in phosphorus. If a soil when analysed is found to be low in all nutrients, one would therefore set out to apply about 50 per cent of the plants' total needs *initially* in the form of a base fertiliser, and supply the balance by seasonal feeding. This method does not necessarily apply

NUTRIENT REQUIREMENT PER PLANT

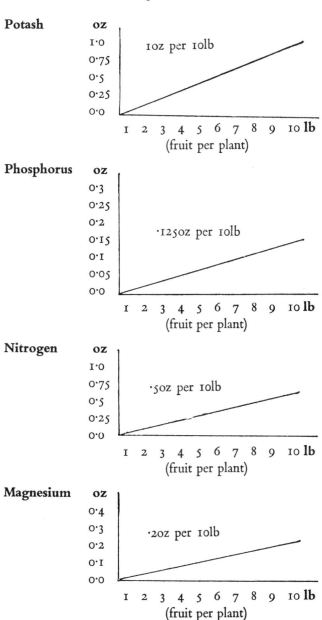

Potash oz

1·0   1oz per 10lb
0·75
0·5
0·25
0·0

1 2 3 4 5 6 7 8 9 10 **lb**
(fruit per plant)

Phosphorus oz

0·3
0·25
0·2
0·15   ·125oz per 10lb
0·1
0·05
0·0

1 2 3 4 5 6 7 8 9 10 **lb**
(fruit per plant)

Nitrogen oz

1·0
0·75   ·5oz per 10lb
0·5
0·25
0·0

1 2 3 4 5 6 7 8 9 10 **lb**
(fruit per plant)

Magnesium oz

0·4
0·3   ·2oz per 10lb
0·2
0·1
0·0

1 2 3 4 5 6 7 8 9 10 **lb**
(fruit per plant)

to soil-less media, it generally being the practice to have soils lower in nutrient status early in the season, and to apply regular applications of liquid fertiliser. The reason for this will be outlined in Chapter 5. Assuming that a tomato base fertiliser containing 8 per cent N, 8 per cent P_2O_5, and 12 per cent K_2O (a standard potash base dressing) is applied at 12oz per square yard, and that there are two tomato plants per square yard with a planned cropping yield of 10lb per plant, the 8 per cent of 12oz nitrogen and phosphorus (P_2O_5) is 1oz each and the 12 per cent of potash (K_2O) works out at $1\frac{1}{2}$oz per square yard; as there are two plants per square yard then the amounts are halved, which reduces them to:

	Amount supplied per plant	Amount required per plant
Nitrogen	$\frac{1}{2}$oz less 50% = $\frac{1}{4}$oz	$\frac{1}{2}$oz
Phosphorus	$\frac{1}{2}$oz less 80% = $\frac{1}{10}$oz	$\frac{1}{8}$oz
Potash	$\frac{3}{4}$oz less 50% = $\frac{3}{8}$oz	1oz

Base dressings are generally not usually applied as heavily as 12oz; 8oz per square yard is a normal dressing. It should also be noted that the percentage nutrients present in fertiliser are quoted as K_2O and P_2O_5, lower figures of potassium (K) and phosphorus (P) actually being available. Roughly only about *half* the crop's needs are supplied by base dressings. Few soils are completely devoid of nutrients, however, and generally speaking the picture is built up as follows:

		Requirements for plant
	Base application 8oz	One-third is met
plus	Reserve nutrients in soil	One-half is met
plus	Top dressing in summer with solid fertilisers, ie 6 or 7 dressings of 1oz per sq yd (roughly equal to total amount of base dressing), or more continuous liquid fertiliser application	Requirements are fully met

Calculating Quantities of Liquid Feeding

There are several ways of calculating the quantities of liquid nutrients required. The simplest example is where 2oz dried blood is mixed in 1gal water. Assuming that one gallon is given to one plant then each plant receives 12 per cent (the nitrogen percentage of dried blood) of 2oz = $\frac{1}{4}$oz. If this quantity were given to several plants, which is much more likely, then the weight given to each plant would be proportional. When 'balanced' liquid fertilisers are applied, things become more complicated, it being important to remember that a percentage of the nutrients contained in each has been given to the plants, not the total quantity of fertiliser (see table on page 163).

Where liquid feeds are bought instead of self-formulated, their analysis should be carefully checked. For simple and rough calculation purposes, if there is one gallon of liquid fertiliser which weighs about 11$\frac{1}{2}$lb and the analysis is 10% N, 10% P_2O_5, and 10% K_2O, then very approximately 1lb of each nutrient will be applied when the whole gallon of liquid fertiliser is used. As volume of liquid fertiliser and not weight is applied (although some liquid fertilisers are sold by weight for dilution), it is necessary to make further calculations. Assuming that one-tenth of the gallon of concentrated liquid fertiliser is used, ie slightly over 1lb, then $\frac{1}{10}$lb of each nutrient is applied. If there is a dilution rate of 1 part concentrated feed to 200 parts (by volume) of water and the whole of this were to be applied, which is 200 parts (25 gallons), then the plants would have received this weight of $\frac{1}{10}$lb of each nutrient, and to calculate the quantities of fertiliser given to each plant it would be necessary to divide by the number of plants.

Obviously gardeners will be working in much smaller quantities of liquid fertiliser, but the procedure is similar, and I would strongly advise that they should think in terms of the total quantity of liquid fertiliser given over a period rather than make calculations on individual plant applications.

The most important issues are (1) the dilution rate, (2) the salt concentration of this dilution (preferably a pC of 2·8–2·9), (3) the balance of nutrients to each other—nitrogen to potash ratio, and (4) the total quantity of fertiliser given. More will be said about this in Chapter 11.

Important note. It should be pointed out that fertiliser percentages are still quoted on P_2O_5 (for phosphorus) and K_2O (for potash) and that the actual quantities of phosphorus (P) and potash (K) available to the plants will be less than the P_2O_5 and K_2O figures stated. This makes accurate working out of the quantities of nutrients supplied still more difficult.

BASIC INGREDIENTS FOR GROWING MEDIA

Loam

Loam is classified largely on the basis of the texture of the mineral particles, and taking particle size as

> sand: 2·0–0·02mm
> silt: 0·02–0·002mm
> clay: less than 0·002mm

the prevailing composition would be in the order of 27 per cent clay, 28 per cent silt and less than 45 per cent sand. Apart from the mineral particle content, the organic matter content of loam is highly important by virtue of its colloidal properties, colloids having great virtue for the promotion of chemical change. The breakdown of the organic matter by micro-organisms results in the formation of humic gums which bind the finer particles together to form the ideal crumb structure, which in effect is a honeycombed collection of soil particles, allowing free passage of air and water. Loam derived from pasture turf, stacked and matured and then put through a shredder, serves as ideal organic matter and has an excellent in-built structure, especially if there has been a great deal of stock feeding on the pasture over the years. Lawns lifted and stacked are also frequently used as sources of loam.

On analysis loams show remarkable variation, ranging from 6–20 per cent organic matter (including moisture), a pH of around 5·5–6·5 (it could be lower or higher) and phosphate and potash levels of low to medium, according to the precise management of the turf. Obviously there is no such thing as a 'standard' loam, and indeed it would be difficult to give precise specifications for loam, as quality varies greatly according to source.

Supplies of good loam, free from pest, disease and weed, are therefore limited, especially in large consistent quantities for commercial and amateur use. Gardeners can often assess the quality of loam by feel. When by squeezing in the hand it is possible to compress it so that it retains shape without stickiness, this indicates a clay soil. At the other end of the scale a sandy soil, particularly one devoid of organic matter, will quickly crumble following hand compression. There is not however the same emphasis placed on loam for tomato culture now as was the case years ago, and this, in many ways, is to be regretted.

The ideal water-air-nutrient balance of a really good loam is something which all plants appreciate, and the tomato is no exception, whether for propagation or full season growing. Yet recent experience with all peat or peat/sand composts seems to raise doubts about the real value of loam, particularly when the cropping performance of plants grown entirely in these media is seen to be invariably of such a high order. The virtues of peat are discussed in the next section, but we may note at this point that the inconsistency of loam (as compared with peat) is its greatest weakness as a growing medium.

Loam used as an ingredient of either seed, potting or growing media, should preferably be sterilised by heat, although chemical sterilisation is also reasonably effective.

Specification for medium loam.
>7–27% clay
>28–50% silt
>less than 52% sand
>6–20% organic matter—generally around 10–12%

Peat

Peat is derived from deposits of organic matter which have been prevented from complete decay by the presence of excess moisture and acidity. *Blanket bogs* develop in areas of fairly high elevation where rainfall is high, while *basin bogs* are developed where moisture is trapped, thereby raising the water table. There are several ways of categorising peat, such as the nature of the plants composing it, the age and depth of the deposit, and whether deposits are pure or contaminated with river or rain-washed soils and silts. Age of deposit and depth of harvesting in relation to age are particularly important as far as horticultural peat is concerned. The degree of decomposition is very important, as this dictates the ability of the peat to remain cellular and thus serve as a porous breathing entity which will encourage vigorous root development, as opposed to merely being a mass of almost humic soggy organic matter which excludes air.

Brown cellular peats feel springy and dry, whereas black humic peats feel soggy and wet. Peats which are too new will still have the remains of plants running through them, and this is not particularly desirable for formulating growing media, nor, at the other end of the scale, is black humic peat by itself desirable, although a proportion of humic peat to provide trace-element supply is now favoured.

One way of classifying peat is by the Van Post scale, which is quite simply based on the colour of the water which can be squeezed out of the peat under a certain pressure. Peat is intrinsically acid in nature, especially if derived largely from sphagnum moss, and pH figures of 3·5–4·3 are usual. In contrast, sedge peats often have a pH of around 6·5 but are not so commonly used in horticulture, as supplies are somewhat limited compared to sphagnum peat, which is available in vast quantities in Britain and in Europe.

It is often puzzling to see the variety of grades of peat which are offered for sale, but a little thought makes it quite clear how they should be used: the coarse fibre-containing peats are for soil improvement and top dressing, whereas the fine granular-textured peats are much sought after for growing media because of their excellent and consistent physical qualities. An average analysis of peat is:

N %	P_2O_5 %	K_2O %
0·7–3·0	0·1–0·2	0·1–0·3

Specification for peat. Brown, granular texture, having a springy feel, with cell walls which are intact so that a porous, breathing, growing medium is provided, capable of storing moisture and nutrients for gradual release to plants.

Sand

Sand largely tends to be graded according to particle size, and while for many building purposes this may be the important issue, for horticulture other criteria must be considered, and of these inertness is probably the most important. Ideally sand should have a pH of around 6·5 and be thoroughly washed, or alternatively naturally 'clean'. Too much calcium, iron, or other mineral ingredient can severely upset the balance of a growing medium especially as sand is used in bulk.

Size gradings are usually stated quite clearly by the quarry concerned, and generally speaking a mixture of fine and coarse sand is more useful for seed sowing and conventional propagation in pots, and for full-season growing media coarser sands are desirable, largely on the grounds of their physical separation and aerating effects. For growing-room activities where capillary watering systems are employed, a coarse sand is certainly advisable to ensure that the compost does not become soggy. The John Innes compost specifies a sand of $\frac{1}{8}$in grist, which is very coarse, and here again aeration and drainage are the main considerations.

It is difficult to be dogmatic, not only on the gradings but even on the necessity of sand in growing media, as excellent results can be obtained by the use of 100 per cent peat composts. Obviously, therefore, there is considerable room for trial and error, although there are certain broad rules which should be followed.

Specification for sand. Coarse, even-particled, chemically inert and clean, with a pH of 6·5–7.

Other Constituents of Growing Media

Apart from loam, peat and sand, other materials are now being used for formulation of growing media, and these include the following:

Vermiculite. An expanded mica, this is an absorbent, light, relatively inert material with a large surface area. Vermiculite is largely used in soil-less mixes, which are based on variable percentages of vermiculite to peat—generally around 75 per cent peat to 25 per cent vermiculite (by bulk), or even a lower proportion of vermiculite.

Perlite. This is a volcanic ash, and again is relatively inert. Remarkably light, it separates components physically, simply because of its large bulk. It is used at the same rate as vermiculite.

Expanded polystyrene and other synthetics. Used in rather similar proportions to vermiculite and perlite, these materials are usually waste from plastics factories and are now being used for compost formulation. While chemically inert, they usually remain a separate entity within the soil, and indeed can remain undecomposed in the soil for an indefinite time.

Basically there are three forms of synthetic—polystyrene foam, polyurethene foam and urea formaldehyde. The first two, while standard enough in specification, suffer the disadvantage of being too cellular in nature, which means quite simply that they are unable to act as a colloidal material in the same way as decomposing peat or rotting turf. Urea formaldehyde foam is, however, a different material, being cottonwool-like and absorbent, and in essence resembling a synthetic peat.

Polystyrene and polyurethene foam, while physically valuable in soils or composts by virtue of their mechanical soil-conditioning powers, have their limitations and are virtually incapable of storing moisture and encouraging chemical change. Synthetics have high insulation properties and therefore impart warmth to a growing medium—and they are of course very light to handle. Experimental work in Germany, America and Britain has shown quite clearly that remarkably good results have been obtained where urea formaldehyde foams are used, not only in growing media, but also for outside crops. Roots have shown excellent development in the area of dispersal of the foam, proving that it gives excellent aeration and growth opportunities. Much research work, however, remains to be done before these materials are commercially marketed in quantity, yet I feel that it will not be long before they are generally used in some spheres of horticulture apart from the small plastic blocks now available.

Waste paper and other materials. Experimental work has been carried out with waste paper and a number of other materials as growing media, but as yet most formulations are still in the experimental stage—especially wood bark in granular form—and none are freely available for sale.

Basic Principles of Soil-less Culture

The variable nature of loam, difficulties of supply and of transport, the possible presence of pests, diseases and weeds, and the problem of sterilisation, are undoubtedly troubles of sufficient magnitude to detract from its popularity as a growing medium. It was to be expected that alternatives to loam would be sought on a wide horticultural front, and pioneer work in this direction was carried out at the University of California. The range of soil-less mixes suggested by this body has formed the basis for most of the research work into soil-less media which has been carried out in Europe, although research has been carried out independently in several countries for many years.

The characteristics of loam have been alluded to previously, and if we understand that loam has a mineral content, organic matter, and a vast spectrum of micro-organisms, it can readily be appreciated that to use loam either by itself as soil comprising a greenhouse border, or as a constituent of compost, is to provide a nutrient-producing 'factory'. This applies not only to the nutrients already present in the loam, but to the loam as a vehicle for the chemical changes necessary before many nutrients can be made available to the plants. Loam has the capacity for storing up these nutrients, well in excess of the plant's needs, until required. Ideally, therefore, given a 'perfect' loam, cultural conditions are excellent.

As a complete alternative, plants can be grown in inert gravel or sand, or for that matter in water, provided that the nutrient requirements of the plant are met by the supply of soluble *and* palatable nutrients. Hydroponics is based upon this conception, and hydroponics is of course perfectly practical under strictly controlled systems of cultivation. Consider cellular peat with all its assets as a water-holding medium, coupled with inert sand to assist with aeration and drainage, and one appreciates the philosophy of soil-less culture. Peat is relatively sterile and so is sand, therefore some nutrients must be supplied in immediately available form, when they are then instantly available to the plants. It should be borne in mind, however, that micro-organisms soon invade the initially sterile medium from the plants in it, and from the atmosphere, thus producing cultural conditions approaching those offered by loam. The time for the breakdown of the peat will of course vary, and perhaps matters should not be unduly complicated by assuming that anything like the conditions of loam culture will occur in the short-term period of cultivation of the normal tomato crop.

More will be said later about soil conditioning for border soils of poor texture, to bring them up to the cultural standard necessary for

COMPOST CHART FOR TOMATOES

Type of compost	Parts of loam (by bulk)	Parts of peat (by bulk)	Parts of sand or grit (by bulk)	Fertiliser quantities per bushel*	Ground lime-stone	Comments
John Innes Seed	2	1	1 (⅛in grist)	1½oz superphosphates	¾oz	For seed sowing only
John Innes No 1 Potting	7	3	2	¼lb JI base (2pt hoof and horn, 2pt superphosphates, 1pt sulphate of potash) by weight	¾oz	For potting young plants
John Innes No 2 Potting	7	3	2	½lb JI Base	1½oz	For full season growing(do not use JI No 3) Use as directed
Levington and others	already formulated					
Soil-less mix		75% / 50% / 100%	25% / 50%	2oz hoof and horn† 1/5oz nitrate of potash 1/5oz sulphate of potash 2oz superphosphates 3oz magnesium limestone ½oz fritted trace elements 2 5 3 A	3oz	For winter and *early* spring; for potting; for seed sowing add only 4–6oz lime and apply liquid feed after germination
				4oz hoof and horn† ¼oz ammonium nitrate 1oz sulphate of potash 2oz superphosphates 3oz magnesium limestone ½oz fritted trace elements 2 5 3 A	3oz	For spring and summer use; for full-season growing
				Alternatively use a complete slow release trace element containing fertiliser (eg Vitax Q4) at 4oz per bushel for potting, and 8oz for full-season growing, or as directed by suppliers	4–8oz usually 6oz	Potting for full-season growing, the higher rate of lime for 75/25 and all peat mixes Can be 3–4oz magnesian and 3–4oz ground limestone

* A bushel measures 22 × 10 × 10 inches. There are approximately 22 bushels in 1 cubic yard. 1 bushel = 1·28 cubic feet.

† Do not add hoof and horn until just before using. Nitroform (38%) may be used in place of hoof and horn at equivalent nitrogen

tomatoes, but it should be remembered that a good loam-less or soil-less medium has an ideal texture from the outset, and this has proved to be highly advantageous. All is not on the credit side, however, as problems certainly exist with loam-less culture, not least of which is the apparent inability to 'soften' or 'buffer' nutrient availability to the plant, something which has both advantages and disadvantages according to the degree of precision exercised.

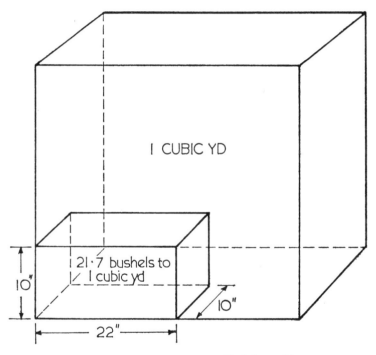

16   *Comparative measurement of bushel quantities*

Compost Mixing

The ingredients for formulating compost should preferably be under cover and *reasonably* dry, although the peat should be moist. Mixing should be carried out on a solid clean floor (making particularly certain that it is not contaminated with weedkiller or other obnoxious chemicals). Measure out the respective quantities in bulk (not necessary where peat is purchased with bulk stated), putting the largest quantity down first, then add the fertiliser, turn once, then add the lime and turn *twice more*. Mixing the fertiliser with a small quantity of dry sand facilitates even distribution. As shrinkage occurs in mixing, ideally the fertiliser should be applied to the final measured bulk, but this can be difficult and labour consuming. After mixing *a suitable period must be allowed, especially with soil-less media, for neutralisation to occur,* 10 *days being the minimum period.*

The keeping qualities of formulated composts vary. Those that contain slow-release forms of nutrients (eg Q4), are more stable than those with 'straight' chemicals, although this is not always the case; it is important, of course, to keep them dry. Experimental work is proceeding constantly on compost formulation and in Ireland, at Auchincruive, and elsewhere has had successful showings with 100 per cent peat compost for full-season growing in bags, troughs, and trench systems of culture.

6

Tomato Propagation

Tomato propagation is a precise task, especially for the early crop, and is best attempted in a well-sited and well-designed glasshouse with adequate heating, preferably automatically controlled. The greenhouse itself should be clean and hygienic, having been carefully washed down during the winter with a good detergent, and the glass cleaned outside immediately prior to use so that maximum light can be transmitted. It is not generally realised how much light is excluded by dirty glass. Proprietary materials can be used for glass cleaning, or alternatively oxalic acid crystals, 1lb to 1 gallon of water, can be sprayed on and then washed off with a hosepipe, preferably when the greenhouse is empty.

Tomatoes can be propagated vegetatively from 4–5in cuttings rooted in a mixture of peat and sand at 65° F (18° C). A supply of cuttings must be available in the first case, which means the over-wintering of parent plants if new young plants are to be produced early enough. Cuttings are useful for later crops, especially to make up shortages. But because of the prevalence of virus disease which would be transmitted by vegetative propagation, seed is therefore the safest and most reliable way of producing tomato plants (see note on page 185).

GERMINATION

Examination of a tomato seed shows that it has a relatively hard coat or testa which effectively protects the embryo from outside influences. Before germination can take place, this seed coat must be softened sufficiently to allow entry of water to trigger off the complicated process of germination. Much seed tends to lapse into what is called a dormancy period, and tomato seeds are no exception, although they are not so temperamental in this direction as many others. It is important that the temperature should be of a sufficiently high level to induce germination, and this varies for the species of plant in question; there must also be a supply of air to allow respiration to take place.

Germination occurs by the production of, first, the radicle from which the root develops and then the plumule which goes up in the air, developing into the shoot. In the case of the tomato the plumule rises in a bent shape. The first two seed leaves develop from the two cotyledons and the testa or coat of the seed should stay below compost level, but is frequently taken up with the seed leaves before falling off, and this increases the risk of virus infection, as the seed leaves may make contact with the *outside* of seed coats. Virus can also be transmitted to other plants by handling, but washing hands (or tools) in a 2 per cent solution of tri-sodium phosphate is an excellent inhibitor.

Once leaves form, photosynthesis takes place and growth proceeds rapidly. What can be seen purely as a shoot and some leaves does not reveal what is going on inside the little plant. What in fact is happening is that flower trusses are being formed at a very early stage, and the result of considerable research has shown quite clearly that the temperature and light levels during this early formative period are critical in respect of the number of flowers, their time of appearance and, perhaps more important, the ability of these flowers to produce fertile pollen.

PROGRAMMING THE CROP

It is essential to decide exactly when seed is to be sown to produce plants which will be at the ideal stage at the right time for planting out. There must be co-ordination between temperature and light levels in the greenhouse, the latter varying regionally. The programmes quoted are therefore subject to modification, but are generally applicable to Britain and countries in similar latitudes. (See page 97 for temperature levels, and for supplementary lighting and growing rooms see page 44. Further propagation details, including propagating temperatures, are given on page 93.)

Early Crop—No Supplementary Lighting

Seed sowing: mid- to end November

Potting-up: end November to early December (8–12 days after sowing)

Spacing: 3 times: preferably in early mid-December (6–8in), early to mid-January (12in) and late January (15–16in) if space allows

Planting: 1st to 3rd week February (according to natural light levels)

Picking first fruit: mid-March to early April

Note that with natural illumination it takes 12–14 weeks *on average* during the winter period from seed sowing to planting, and a further 6–7 weeks until first fruit is ripe. Note also that with later crops this time is considerably reduced. There is bound to be a considerable variation according to the area and the level of natural illumination. The better the light the shorter the propagating period. The earliest crops in Britain are normally produced in the south, and gradually become later as one journeys north. It is essential to remember that very early crops should only be attempted in good light areas and where cultural facilities are excellent.

Early Crop—Supplementary Lighting

Seed sowing: late November to early December

Potting-up: early December to late December (after 8–12 days); light treatment then given for 17–21 days according to rate of growth or until plants require spacing.

Spacing: this is carried out as for the unlit crop—generally 2–3 times

Planting: again variable, but generally from late January until early February, the earlier planting being achieved because of benefits derived from supplementary lighting

Picking first fruit: varies from mid-March until mid-April; the net effect of supplementary lighting will greatly depend on the precise area and type of season

Note that on average it takes 9–10 weeks from seed sowing to planting with supplementary lighting.

Early Crop—Growing-room Propagation

Seed sowing: mid-December

Potting-up: late December (after 8–12 days), when light treatment is given in the growing room for 14–21 days (12–16 hours daily) until first truss is initiated (see page 88); small pots are used during light treatment, the plants being potted into 4¼in pots afterwards

Spacing: into propagation house at 6 × 6in or thereabouts, when some supplementary lighting is then valuable for 12–16 hours daily, for 7–10 days; space thereafter as necessary

Planting: from late January till early February

Picking first fruit: early March to early April (once again it must be stressed that the natural light level in any area is a vital factor in ripening bottom trusses of fruit)

Note that from seed sowing to planting can take as little as 7–8 weeks under ideal conditions, which is a saving of 4–5 weeks on natural propagation.

Second Early Crop—Natural Propagation

Seed sowing: early to mid-December

Potting-up: late December to early January (after 8–12 days)

Spacing: as for early crops

Planting: end February to mid-March

Picking first fruit: late April to early May

Note that propagating time varies greatly according to natural light level in the district but is generally 10–12 weeks. If lights are used, shortening of propagation period is likely.

Mid-season Crop

Seed sowing: late December to early January

Potting-up: mid-January (8–12 days later)

Spacing: as before

Planting: mid-March to early April

Picking first fruit: early to mid-May
Note that once again propagating time varies greatly according to the natural light level of the area, but is generally 9–10 weeks.

Late Crop

(applies to crops grown in moderately heated greenhouses)
Seed sowing: from mid-February
Potting-up: late February
Spacing: generally on 2 occasions
Planting: mid to late April
Picking first fruit: June
Note that the propagating period from sowing to planting is now shortened to 8–9 weeks.

Later or Cold Crops

(including crop to follow on after an early crop)
Propagating with heat, but cold cropping in greenhouses (though in the north *mild* heat in the greenhouse is always advantageous, especially at night to avoid high humidity).
Seed sowing: early to mid-March (in heat) until May for very late crops
Potting-up: late March to early April (in heat for early period) until May for very late crops when potting can be into large pots
Spacing: as for late crop
Planting: early May to June (if standing outdoors until 'planting' time)
Picking first fruit: July to September
Note that the propagation period has now shortened to around 5–7 weeks, perhaps 8 in colder areas.

SOWING AND POTTING

Sowing Procedure

¼oz of seed will produce approximately 1,000 plants. When sowing on a large scale use *clean* seed trays of plastic, polystyrene, or wood, each measuring approximately 14 × 9 × 2 inches. On a small scale clean pots or seed pans should be used. A seed tray will take 250–350 seeds and a 6in pot about 60 seeds. Seed can also be sown individually in pots.

John Innes seed, all-peat, or peat/sand mixes (see page 84) are suitably warmed, put into the receptacles, struck off level and then pressed down to within a quarter of an inch of the top. A rich compost is not desirable for seed sowing, and for sowing direct into a pot it is safer to use liquid feeding once the young plants start showing rapid growth. One major problem with seed sowing is the risk that ammonia damage will cause damping-off, and this can occur when sterilised John Innes compost has been used. It is less of a problem in soil-less mixes.

The seed is sown thinly (about 2–3 per square inch) either broadcast or in ⅛in deep drills made with a piece of cardboard. It is then covered *lightly* by riddling on some compost or by closing the drills, then pressed in fairly firmly, with the object of keeping the seed testa below the compost level to avoid virus infection (see also page 88).

Alternatively seed can be spaced out at around 48 per box or thereabouts, or directly into small pots, a procedure which avoids checks during pricking off; there is (according to various research findings) some virtue in this and it is easier if pelleted seed is used.

Sowing Preparatory to Grafting

It is useful to have plants at a wider spacing preparatory to grafting, and spaced sowing at 24 seeds per tray is frequently practised, although the seed can be sown closer and pricked out into boxes at 24 per seed tray thereafter. One variety and one root stock seed can be sown in individual pots. Note that it is generally advisable to sow the seeds of the root stock some 7–10 days *before* the variety owing to its slower germination.

Germination Techniques

Tomato seeds are light-hard, which means that they germinate better in darkness. After sowing seed, lightly water with a *fine* rose, taking care not to wash off the covering. Immersion of the base of smaller receptacles can be a useful technique, using water that is 'greenhouse warm'. If the receptacles are either stacked (in the case of seed trays) or covered with a sheet of paper or black polythene, moisture is conserved on the surface, assisting germination. Much use is made commercially of germinating cabinets, which are highly insulated shelved cupboards where humidity is high and temperatures can be accurately controlled. As soon as germination occurs reasonably uniformly through the seeds (there will always be a proportion which do not germinate as quickly as the others) the boxes are unshaded, the covering removed, and the seedlings given full light. Drying out must *never* occur during the actual germination period and additional watering may be necessary.

Despite considerable controversy about germination temperatures for tomatoes, it has been generally agreed by plant physiologists that a 65° F (18° C) day and night temperature is generally acceptable. Higher temperatures will induce a higher percentage of rogues (short-jointed squat plants with an even 'true leaf' development, but which will remain unproductive). The production of rogues, although influenced by environmental conditions, is thought to be genetical. Gardeners who fear that they may have missed rogues during pricking-off will have little difficulty in noting their squat short-jointed, side-shooting habit at a later stage. Although they may look healthy and vigorous, they should *not* be planted.

Containers for Potting

Within 8–12 days of sowing there should be a sufficiently high number of seedlings for potting. There is some differences of opinion concerning the virtues of early pricking off, especially as some authorities now believe that the first trusses are laid down at an earlier stage than was thought to be the case and that early pricking-off could coincide with this critical period. In the absence of fully confirmatory evidence, however, the normal process of early pricking-off is advised.

Research has shown quite clearly the virtues of pots that are large enough to contain the plants for a longer period, this particularly being the case with the earliest raised plants which are not planted until flowers are showing. A pot size of 4¼in is generally recommended, and pots may be either of plastic, treated paper, polythene, peat or clay. Soil blocks may also be used, provided that these are large enough to sustain growth, which generally means a large block size of approximately 3¼in. Soil-less materials are also available for block making. Blocks are better contained in boxes or trays to avoid any collapse which may occur, especially when wet.

Dark-coloured pots have been shown to absorb solar radiation better than lighter coloured ones. Plastic pots are also warmer, because they do not lose heat by evaporation.

All pots must be scrupulously clean, especially used clay pots which should be soaked in a solution of formaldehyde (1 in 49) several weeks prior to using, or alternatively subjected to steaming. The problem of hygiene, coupled with the extra watering necessary, has tended to make clay pots less attractive, but they are nevertheless capable of producing excellent plants.

Peat pots contained in trays on a shallow layer of peat are extremely useful and have a low water requirement. They should always be kept on the dry side and further apart than other types. Hexagonal paper pots have the advantage of taking up less room, and they can be fitted closer together for brief periods, which is a distinct advantage for light treatment where space is important.

Compost for Potting

There is considerable variation in compost type for potting tomatoes and reference to Chapter 5 will show that the composts used can be (1) John Innes No 1 Potting, (2) soil-less mixes in various proportions, (3) Levington, (4) other proprietary or self-formulated types. Much modification has taken place in recent years, not only in the physical nature of the ingredients but in the fertiliser content. Soil-less mixes tend to give softer plants than those grown in John Innes compost. The further north one travels, the smaller the nitrogen requirement in the compost, particularly for early raising, the nitrogen levels appearing to matter less for later raising when light values are higher. Porosity of the compost is important, especially when plants are raised in growing rooms and watered by capillary methods, for which a very rough grit is used in lieu of the more usual mixture of fine and coarse sand.

It is most important to avoid overloading potting media with fertiliser, for this raises the salt content above the acceptable level of pC 3: lower pC figures denote high salt content.

Potting Procedure (where necessary)

The young plants are teased carefully out of the compost with a clean tally (*not* pulled out, breaking the root). Each seedling is held by the leaf, with the finger and thumb; avoid handling the stem for this can result in damage to the stem hairs or the actual tissue. Two courses of action can now be employed. The first is to have the pots ready and half filled with the clean *warmed-up* compost, and holding the seedling in the centre of the pot with one hand, gently press more compost into the pot round the seedling with the other, firming and levelling with the fingers. The other method is to fill pots to the brim loosely with compost, then with a finger or dibber make a deep hole in the centre of the pot and carefully insert the seedling, so that the seed leaves are about one inch above the top of the compost. The compost is again firmed with the fingers. In both cases the pot is given a sharp tap on the bottom to level up and firm the compost neatly. In the case of soil blocks seedlings should be very carefully inserted, taking care not to break or bend the root when filling the indent with compost. A light watering is now advisable unless the compost was fairly moist in the first case. (Note use of tri-sodium phosphate, see page 88.)

Plants intended for grafting are generally not potted, but put into boxes (12–24 per box, unless of course they have been space sown in the first case). The procedure is exactly the same as for the second method above: the seed boxes are filled and levelled off, the seedling is inserted with a dibber, firmed and finally watered.

CARE OF YOUNG PLANTS

Good light, the correct temperature level, and careful watering and feeding are essential if healthy young plants are to be raised. There must also be avoidance of any irregularities that result in checked plants and render them susceptible to parasitic fungal ailments.

Natural light levels must be good for healthy growth, although much can be done to ensure that plants receive all available light by placing them sensibly. With an east-west propagating house this is invariably on the bench running along the south side. An open slatted bench with warm pipes 9–12in below the bench is also ideal, although solid benches can give excellent results, and may in fact be a necessity where electrically warmed benches are being used.

Temperature Levels

Where the precise control of temperature is possible, it is necessary to decide on the correct day and night temperatures to give, and there is some difference of opinion on this between northern and southern research stations, and also between British and European research establishments. The recent establishment of a Glasshouse Research Station in Scotland (at Auchincruive, Ayrshire) should now add further to the fund of available knowledge.

It will be appreciated, from the discussion in Chapter 4 of the physiological aspects of growth, that when light and temperature levels are high photosynthesis can generally proceed at an optimum rate in the absence of any inhibiting factors (such as shortage of CO_2, moisture, nutrients, etc). Food manufactured by photosynthetic activity is required by the growing plant, and where photosynthesis is depressed because of low temperatures and poor light some general debility is unavoidable. This may not necessarily be visually apparent, but would affect the formative period of fruit trusses, which are laid down in the plant before they are visible.

Higher day temperatures, with sufficient light intensity, generally increase the rate of growth but reduce the number of flowers on the bottom truss, and this affects the potential number of fruits. High daytime temperatures do in fact induce earlier ripening of the bottom truss, perhaps quite understandable if one assumes that chemical activity proceeds more quickly.

Lower daytime temperatures increase the number of flowers on the bottom truss (and of course the potential number of fruits) but can delay ripening, and there may also be some effect on the viability of the pollen, resulting in poor setting (fertilisation).

The effect of lowering night temperatures is not quite clear, although it is generally accepted that there must be a drop in night temperatures to reduce the respiration of the plant, and so avoid dissipating the available supplies of carbohydrates and protein. Night temperatures must be 'balanced' to daytime temperatures to avoid either an over-supply of carbohydrates (resulting in squat plants) when night temperatures are extremely low, or long leggy plants when night temperatures are too high.

Good light levels are an asset irrespective of temperature levels, although where light levels are poor in high daytime temperatures there will again be dissipation exhibited by leggy growth, as the plant is unable to manufacture carbohydrates to keep pace with growth rate.

The net result of very low temperatures will be the production of hard 'blue' plants (see page 59) which eventually affects fruit production. The possible onset of fungal diseases later in the season due to growth checks cannot be overlooked.

The following tables show temperatures quoted by the respective research authorities for the south and north of England respectively, the northern recommendations applying to the *whole* of Scotland. Notes on stages 1–4 will be found overleaf.

For Southern Growers

Stage	Night		Positive Day		Ventilation commences at	
	°F	°C	°F	°C	°F	°C
1	58	14	64	18	74	23
2	60	16	64	18	74	23
3	62	17	65	18	75	24
4	62	17	64	18	70	21

For Northern Growers

	°F	°C	°F	°C	°F	°C
1	56	13	70	21	74	23
2	56	13	68	20	74	23
3	56	13	68	20	74	23
4	60	16	68	20	70	21

Stage 1: from pricking out the seedlings to the visible appearance of the flower buds in the first truss.

Stage 2: from the first truss flower buds to anthesis (for this purpose the opening of the first flower in that truss).

Stage 3: with the first blooms bursting into colour, the plants are planted out into borders or still in pots and the period extends to four weeks after picking the first ripe fruit.

Stage 4: to end of cropping.

Temperatures refer to those accurately recorded by a screened thermometer (or at least shaded from hot sun).

For the period of propagation we are concerned with stages (1) and (2). There must usually be some compromise, especially as it is exceedingly doubtful if the degree of precision possible in a highly sophisticated commercial glasshouse unit can be emulated in all spheres of culture, particularly amateur gardening. A fair target, therefore, is about 56–58° F (14° C) at night and 63–65° F (18° C) by day, as far as this is possible, lowering the day temperature a little in very dull weather. Normally, without sophisticated instruments, the temperature level tends to drop at night and rise automatically during the day, especially if there is solar radiation. Note that ventilation should take place at 70–74° F (21–23° C). A maximum and minimum thermometer is a good investment for checking temperatures.

Different varieties of tomato respond differently to varying temperature ranges, but it is impossible to be very precise about this at the present stage of research. Research work is proceeding on the value of a temporary drop of temperature, short periods of high temperature, and other treatments.

Spacing, Watering, Feeding and Supporting of Young Plants

Spacing. Young plants must be spaced out as soon as their development demands it, generally three weeks or so after potting and again in another three weeks, and finally a short time before planting out. When plants are congested their leaves are not fully exposed to the light and photosynthesis is reduced, causing plants to become drawn, leggy and etiolated, which raises the height of, and may debilitate, the bottom truss.

Watering. There can be no precise ruling on the quantities of water required by young plants. Free-draining compost with a lot of grit requires more water than an all-peat mix. Clay pots dry quickly, paper, plastic and peat less so, indeed there can often be a danger of over-watering peat pots, which should not be kept too close together as they become soggy. Soil blocks are best kept close together initially to avoid drying out, as once a soil block is dry it can be very difficult to re-wet it. Continual observation is the only way to gauge the plant's needs, applying water carefully with a long-necked watering-can, or a slow-running hose if a large number of plants are involved, taking pains to avoid saturation or, at the other end of the scale, wilting of the plants. Research has shown that it apparently makes little difference whether water is at greenhouse temperature or straight from the water supply, although many gardeners prefer to use water from a tank in the greenhouse.

Feeding. There is some controversy concerning the need to feed plants in pots. Plants in an adequately sized pot in a good John Innes No 1 Potting Compost could well carry right through to the planting-out stage with no debility. Research workers, however, claim that in this case there will have been a shortage of nutrients causing later debility, which will show up as small fruits on the first or second truss. On balance I feel it is advisable to feed every 10–14 days. Plants in soil-less media should certainly be constantly fed, possibly alternating feeding with plain water to avoid salt concentration build-up, using high potash feeding initially (see page 163) at 1:200, or proprietary feeds at recommended dilutions. Foliar feeding is claimed to be highly advantageous at this stage and it certainly avoids excess salt concentration in the pots. Plants which are propagated later and develop quickly, frequently do not seem to require so much feeding, although I feel that feeding is always desirable with soil-less mixes.

Supporting. Plants allowed to flower before planting out will usually require support, generally with a split cane and fillis.

GUIDE TO YOUNG PLANT RAISING

The visual appearance of a plant is the only real guide the grower or gardener has regarding its well being, and the following are the main points to look for:

Symptoms	Cause	Treatment
Pale green plants with long internodes, despite regular feeding and good temperature control	Lack of light combined with too high temperatures	Drop day *and* night temperatures slightly
Yellowing lower leaves, slow growth and poor colour despite regular feeding	Excess watering excluding air from compost which inhibits nutrient uptake	Reduce watering and give an organic stimulant such as dried blood at 2oz/gal
Spindly, leggy growth despite good colour; burning and shrivelling of leaf tips evident	Excess heat, generally during the day, caused by lack of, or insufficient, ventilation	Ventilate at 70°F (21°C) and damp down frequently; if absent from home all day in hot weather and there is no automatic ventilation, leave vents open to be on the safe side
Dark blue colouration of leaves, especially lower ones, coupled with stocky hard growth	Excess cold in whole greenhouse, or in one area affected by draughts	Check operation of heating, especially at night, using a maximum and minimum thermometer; put up polythene curtains if necessary to restrict draughts, especially in the region of badly fitting doors
Very dark green plants with excess curling of leaves, yet plants do not grow quickly; shrivelling of leaf edges,	Excess fertiliser or liquid feed being applied, resulting in high salt concentra-	Reduce fertiliser or liquid feed applications giving a period of plain water to flush out salts

Symptoms	Cause	Treatment
especially growing tip	tion in the grow-ing medium	
Seed leaves dead and plants look pale or stunted	One or other of the above. The well-being of the seed leaves is an excellent barometer on the well-being of the plant	Make general check on grow-ing conditions, and use a liquid fertiliser at the correct dilution
Leaf mottling, 'blind-ness' of apical point and growth depression	Unbalanced fertiliser levels or temporary shortage of trace elements	Check feed concentration; minor irregularities usually correct themselves; possibility of weed-killer contamination should not be dismissed
Wilting; general debility or distortion	Fungal or virus disease, or pest attack	See Chapter 12 for full account of pests and diseases—symp-toms, preventive and curative controls; in general terms checks to growth result in development of an excess of carbohydrates which can readily be plundered by weak pathogens

GRAFTING TOMATO PLANTS

The use of root stocks is common practice in many spheres of horticul-ture and involves exploiting the virtues of the root stocks for the pur-pose of overcoming inherent weaknesses in the more productive plant grafted on to it. Thus weak varieties of rose are grafted or budded on to vigorous root stocks, strong varieties of apple are grafted on to weaker root stocks, and so on. Grafting usually shortens the vegetative propagation period. Tomato grafting is intended to utilise the inherent resistance of the root stock to various maladies which the tomato fruit-ing variety itself would be unable to resist. It is therefore possible by this means to grow tomatoes in soil known to be contaminated with certain troubles without resorting to heat or chemical sterilisation, perhaps where sterilisation is not practical for some reason—a situation quite common in both professional and amateur gardening circles.

Root Stock Seed

The root stocks used are F_1 hybrids (which entails raising under con-trolled conditions) and they are offered by most seedsmen as follows (see Chapter 12 for details of diseases):

KN: corky root and nematodes resistance (root-knot eelworm, *not* potato eelworm, although in practice there seems to be some resistance to both).

KVF: corky root, verticillium and fusarium wilt resistance.

KVNF: corky root, verticillium, nematodes and fusarium wilt resistance.

Note: certain resistances are now being bred into tomato varieties—see Variety List in Chapter 14.

Root stock seed generally has a harder testa than the fruiting tomato variety and may take longer to germinate, in which case it should be sown (as mentioned on page 92) some 7 to 10 days earlier. Prick off both root stock and fruiting varieties into trays (or pots) at the density of 12–24 per tray or box. Alternatively sow one seed of each (root stock and variety) in a pot. Plants should be given plenty of light to en-courage strong growth, as grafting can be difficult if they are soft and leggy.

Grafting Procedure

The actual grafting procedure is carried out when plants of both root-ing and fruiting varieties are about the same height, round about 4–6in tall, with a stem thickness of ⅛in, although it is frequently delayed until the plants are taller (this is wise if the grafter is inexperienced). One seedling of both root stock and fruiting variety is then carefully lifted, and using a sharp clean razor blade, a *downward* slanting cut is made in the root stock $\frac{1}{2}$–$\frac{3}{4}$in long, a corresponding *upward* slanting cut then being made in the fruiting variety. The cut should be made immediately above the area of the seed leaves (which should be removed), taking care not to cut into the stem more than halfway across its diameter, to avoid undue weakening. The two tongues of the cuts are carefully fitted together by one person, while another wraps strips of transparent adhesive cellophane or lead, 1in wide, round the stems at the point of the cuts. General procedure is the same if direct sown in pots.

The top leaves of the root stock are then removed and the plants planted up carefully in pots of $3\frac{1}{2}$–$4\frac{1}{4}$in in diameter. The pots can be closely spaced, usually on an open bench, and they can, if necessary, be draped with polythene to ensure proper humidity while under normal greenhouse conditions; and provided the plants are kept wet but not over-watered, union of the cuts usually occurs fairly quickly, as can readily be seen through transparent wrapping. Temperature levels are the same as those quoted on page 97, and feeding is carried out if necessary.

At planting time it is generally recommended that the fruiting variety root is removed immediately *below* the graft in order to avoid the possible later transmission of fungal vascular infection to which the fruiting plant root is susceptible, but personal experience shows that a cultural check is caused by this procedure. In fact the fruiting plant root generally withers, and is then unlikely to transmit vascular diseases. In any case the remaining upper portion of the root stock should be cleanly removed.

One vital point on the technique of grafting is hygiene during the grafting procedure; it is relatively simple to infect the cuts with disease unless the blade is dipped frequently in a solution of 2 per cent trisodium phosphate.

While grafted plants are slow to establish when planted, they develop considerable vigour which invariably continues well into the autumn.

Note: The whole matter of tomato grafting must be given careful consideration with regard to the resistances now being introduced into tomato varieties. The success of certain alternative cultural systems also throws doubt on the economic viability of grafting.

7

An Examination of Cultural Methods

The actual physical growth of a tomato plant from the time of seed germination to its ultimate demise could be said, botanically speaking, to follow a fairly consistent pattern. Yet there are so many variables to take into account and so many ways of achieving the same results that confusion inevitably arises in the mind of the gardener or grower. There can obviously be no 'best' way of growing tomatoes under all the differing conditions, nor indeed are the results entirely consistent, owing to differences in the weather from season to season, or the progressive onset of disease under particular cultural conditions.

I feel that in previous chapters of this book there has been a sufficiently detailed account of the different circumstances and situations under which tomatoes may be grown for the grower to be able to gauge the level of success he is likely to achieve. Under good natural light conditions, with a well-sited greenhouse of good design, with adequate ventilation and a good level of heating, and where the correct environment can be maintained and supplies of water and nutrients controlled to fine limits, then one would expect, given a good pest- and disease-free medium in which to grow the plants, that success is inevitable. Yet to dwell completely in the realms of such a tomato-growing Utopia would, I feel, be unrealistic, certainly for the greater proportion of those who read this book. It is certainly commendable to strive for the ideal, yet in the course of a great many years' advisory work (combined with the culture of tomatoes on my own account), I can seldom recall a situation where every cultural requirement to the present level of knowledge was completely and absolutely fulfilled.

With tomato culture, perhaps more than with the culture of any other crop, there is always something new to learn or something new to try. The very fact that so many cultural methods exist proves this point adequately.

GROWING METHODS 'BALANCE SHEET'

Conventional Border Culture

The tomatoes are grown in the soil forming the 'floor' or border of the greenhouse.

Credit	Debit
A stable water regime related to water tables of the area. Less water loss than from containers or bales. A border retains moisture effectively, an advantage during absence at business or on holiday.	During wet weather water may seep into the border, causing chilling and excluding air, this being important in the early months of the year. It is not always possible to control accurately the quantity of moisture available. Soil is slow to warm up, frequently resulting in planting checks, with serious long-term repercussions (see page 177). Conventional culture without sterilisation (and even with it in certain cases) can result in problems due to the build-up of pests, such as potato eelworm, and diseases (see page 173). Accumulation of salts and plant acids will build up in the soil. Excess vigour of plants can be difficult to control, especially in freshly sterilised soil-containing media. Re-soiling is laborious and in many cases it may be difficult to obtain reliable soil entirely free from weeds, diseases and pests.

Border Culture Using Grafted Plants (see page 101)

Credit	Debit
The same advantages as are mentioned in the entry relating to conventional border culture, coupled with the ability of the existing root stock (because of their inbred resistance) to grow well despite the presence of several serious maladies, thus avoiding the need to sterilise the soil. Some of these resistances have now been conferred to fruiting varieties grown normally on their own roots. Plants maintain good vigour right to the end of the season.	The technique of grafting can be tedious and if the plants are to be bought they can be difficult and costly to obtain. Specific inherent resistances have a bad habit of breaking down, but the plant breeder is usually able to take the necessary action. Virus disease may be spread by the grafting process. Fruit is usually slightly later in ripening. Vigorous roots contact deep-seated virus infection in soil.

Ring Culture

Plants are grown in compost in 9in bottomless bituminised paper pots or other similar-sized containers, which are planted on a 4–6in layer of porous and inert material such as grit, weathered ashes or gravel.

Credit	Debit
Only relatively small quantities of pest- and disease-free growing media are required annually, ensuring a good start. The growing medium warms up quickly as it is freely exposed to both heat from the sun and artificial heat (it also cools quickly). Excellent control of vigour can be exercised by early root restriction. This results generally in fruit production of a high order. Deep-seated pest and disease problems are of no consequence.	Watering and feeding must be constant and carried out with considerable precision. Water loss is very high, especially in hot weather, and early drying out can have serious consequences, especially later in the season, as root damage may ensue. Setting aside a section of the greenhouse for ring culture tends to limit its use for crops such as chrysanthemums, although they can be grown in ring culture. Salt build-up (fertiliser residues) can be a serious problem in the limited quantity of growing media.

Straw Bale Culture

Plants are grown in a ridge of growing medium on straw bales which have been induced to 'heat' by chemical treatment.

Credit	Debit
Warmth at the roots is ensured, resulting in excellent development. As with ring culture, a 'clean' start is ensured. CO_2 enrichment by decomposition of the straw is beneficial to plants.	Vast quantities of water are essential, both initially to soak the bales, and throughout the season. Bales are heavy and bulky, taking up a lot of room, although sections of bales can be used. Vigour control can be difficult as root development is uninhibited and uncontrollable supplies of nitrogen are released from the decomposing straw. Deficiency of nitrogen can occur, also pests.

Container Culture

1. *Polythene bags.* Plants are grown in soil-less media in large black polythene bags set in a layer of polythene, which acts as a barrier to pests and diseases, or on layers of free-draining aggregate.
2. *Peat mattress system.* 9in whalehide pots are set in a trough of polythene on a 2–3in layer of peat.
3. *Various containers.* Bolsters (polythene sacks with the tops opened up); buckets (with drainage holes); boxes, trays or troughs constructed of polythene or expanded metal.

Credit	Debit
A pest- and disease-free start in relatively small quantities of growing media. Root growth and vigour are usually excellent.	Watering and feeding require considerable precision. Surplus water shed from bags can result in soggy soil and raise the humidity. A salt problem can frequently occur in the media, resulting in 'black bottoms' (see page 197) and other troubles.

SOME ASPECTS OF WATER APPLICATION IN TOMATO CULTURE

Water is the life blood of all plants, and when they are grown under glass the only source available to them apart from reserves in the soil is what is applied by the grower or gardener, though a certain amount of water can travel laterally from outside sources and there can be seepage from moisture shed from the roof of the greenhouse. The application of water is best considered under different heads.

Winter Flooding (where border culture is practised)

The need to flood out surplus fertilisers and top up the moisture reserves in the soil is frequently emphasised in connection with border culture of tomatoes. It is usually done some weeks prior to planting, and where steam sterilisation has been carried out, it follows immediately on this to wash out surplus ammonia. Research experience has shown, however, that apart from chilling the border, excess quantities of water may do much harm by causing damage to soil structure. It is now usual in commercial circles to have the soil analysed to ascertain the soluble salt content, and to use this as a guide as to whether or not flooding is necessary. Generally speaking, it is advisable to give the borders a reasonably heavy watering two weeks or so prior to planting, to avoid their drying out. Where soil analysis shows the need for a thorough flooding, heavy soils will require much more than light soils, the following table being a guide:

Area of border	Light soil	Heavy soil
10sq yd	40 gallons	80 gallons
20 ,,	80 ,,	160 ,,
30 ,,	120 ,,	240 ,,
40 ,,	160 ,,	320 ,,
80 ,,	320 ,,	640 ,,

For larger areas multiply accordingly. Where soils merely need pre-planting watering, a quarter of the quantities stated should suffice.

Flooding should always be carried out in 'mist' form with either a rose on the end of a hosepipe or spraylines, to keep damage to soil structure to a minimum. The amount of water delivered can be measured either by meter or, on a smaller scale, by sample collection in a bucket or other receptacle where spraylines are being used. For example, if a square biscuit tin of 1sq ft surface area is put down and collects 1 gallon of water in 2 minutes, then provided the distribution of water from the spraylines is even over the whole area, a brief calculation will quickly determine the total quantity delivered.

Commercial growers may decide to have the salt content of their soil checked after flooding to see that it has been reduced to a 'safe' level.

Hygiene

Washing down houses inside and out has previously been referred to, and it is a practice which requires water of suitable pressure delivered from the end of a hose so that it can be directed at the glass and into corners. Washing down must fit in conveniently with the cropping plans for the greenhouse. Obviously it should be carried out *before* sterilisation. Prior to washing down, a mild detergent can be brushed on to the glass structure with a long-handled brush.

Damping-down and Pollination (see also page 52)

Water in fine mist form is used for the purpose of raising humidity in the greenhouse so that pollen grains do not dry out and prevent 'setting'. One effect of too dry an atmosphere is 'dry set' (see page 198). Damping-down raises the humidity and also reduces the transpiration of the plant, and this is very important during the establishment period of the plants.

The actual shaking of the plants for the purpose of moving the pollen from anther to stigma is generally carried out either with sprinkler lines or, more usual on a smaller scale, with hosepipes fitted with a rose.

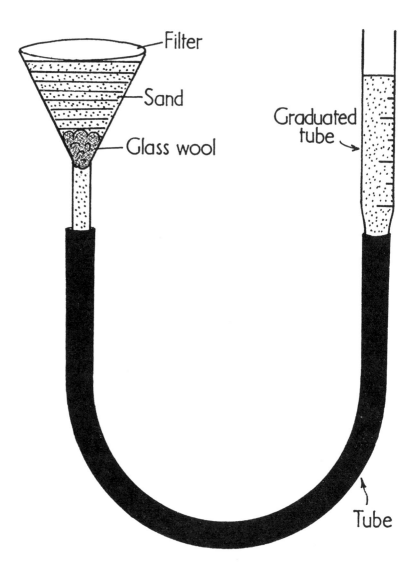

17   *The Jones Rothwell evaporimeter*

When plants are young they can readily be watered overhead, their water requirements being small; this serves also for damping-down. When plants are taller, however, damping-down is then carried out briefly for a few minutes, generally at mid-morning on a sunny day, the vents being shut to raise the humidity for between 30 and 45 minutes. The actual watering of established plants, on the other hand, is carried out according to the plants' needs, estimated either by visual assessment or by other more precise methods now in use.

Assessing the Water Requirements of Tomato Plants

While visual assessment will obviously suffice for a small number of plants, it is desirable to use more accurate ways of determining water needs for tomatoes grown on a larger scale. Tensiometers, for example, record the osmotic pressure of the soil which can be related to a water-requirement table, or more simply state whether the soil is wet, medium or dry. I would think that tensiometers are effective in soil-less media also.

A simple device which depends not on the osmotic pressure of the soil or growing media but on the actual evaporation of the plants is the Jones Rothwell evaporimeter. Composed merely of a graduated tube and sand-filled funnel connected by rubber hose, this equipment can be purchased and put into use for a little over £1. It is placed at the top-most level of the plants and the water requirement is calculated by reference to tables supplied with the evaporimeter (see fig 17).

Another way of assessing water needs is to refer to tables used initially by the Fairfield Experimental Station in Lancashire. A healthy tomato plant will, on average, take up about 22 gallons of water in the full growing season (excluding propagation). This is the quantity required by the plant and does not take into account drainage loss and surface evaporation. The figures shown on next page apply to plants 3ft tall; adjustments must be made for smaller plants.

The requirement ranges from 1¾ pints per week to nearly 3 gallons, a fair *average* being 2 gallons per week. Calculating on the basis of the period between mid-April and mid-September, ie 22 weeks at 2 gallons per week giving a total of 44 gallons, it can be seen what percentage of water is lost by drainage and surface evaporation, especially as more water than this may have to be given. Plants on ring culture could in fact require considerably more water, owing to the porous nature of the aggregate. The same is true of other cultural systems, especially straw bale or polythene bags.

Weather pattern	Water requirement by plant for full 24hr day
Very dull—cloudy and dull most of the day	$\frac{1}{4}$–$\frac{1}{2}$ pint
Dull—overcast most of the day	$\frac{1}{2}$–$\frac{3}{4}$,,
Fairly sunny—cloudy with bright periods	$1\frac{1}{4}$–$1\frac{1}{2}$ pints
Sunny—only occasional cloud	2–$2\frac{1}{4}$,,
Very sunny—sky clear and sunny all day	3–$3\frac{1}{4}$,,

8

Pre-planting Procedures

It has earlier been agreed (see page 104) that the development of the tomato plant from seed to its ultimate and continued production of fruit, follows throughout the season, physiologically speaking, a basically similar pattern no matter what the system of culture or district. This of course can be said about any plant; it is the ability to maximise production, both in respect of total weight and of quality, which brings the more sophisticated aspects of culture into play.

While there may be 'secrets' involved in tomato culture which will coax the plant to give of its best, I feel that it is more important to consider tomato growing as a reasonably precise science.

SELECTION OF CULTURAL METHOD

The decision to grow tomatoes by one or other of the methods previously alluded to need not necessarily entail great deliberation. There should first of all be a critical review of last season's results and, if possible at the end of the season, a close examination of the spent plants, especially their root systems. Detailed pathological and entomological guidance is given in Chapter 12, but the following is a general approach to the subject.

Where plants have cropped well and are removed with a fairly extensive amount of healthy looking white root which is still more or less physically intact, then it can be assumed that soil conditions have been good and there has been little incidence of pests or disease. This sort of situation is common in a 'new' soil for the first year of cropping.

Where the plants have cropped fairly well, yet only a small portion of their root system remains and most of this breaks away on lifting, it can be taken as reasonable evidence of the presence of soil-borne maladies, especially if the remaining root is brown and rotted. This indicates the probable build-up of these troubles to proportions that will affect the ensuing crop seriously, unless evasive action is taken.

Complete collapse of a high proportion of the plants during the season is usually indicative of serious soil-borne troubles—unless only stems or leaves of plants are attacked—demanding sterilisation, re-soiling, or alternative cultural methods (see pages 105–7). Whether or not the soil was new or sterilised before cropping, the presence of troubles shows that pests and diseases were involved and this is the important issue when deciding on cultural methods for the next crop. However, despite the broad terms of cultural method selection outlined above, it is of course possible to adopt ring culture or go on to containers or troughs under all circumstances for reasons other than pest or disease incidence. There may in fact be no suitable soil in the greenhouse border, or the control on vigour made possible by container growing may be desired.

PRELIMINARY CONSIDERATIONS FOR BORDER CULTIVATION

The use of the greenhouse border or 'floor' is still the most common form of culture, especially where the greenhouse is fairly large. The physical issues such as drainage, even soil depth, lack of subsoil pan etc, are important, and it is well worth taking the time and trouble to dig a few inspection holes to check on the subsoil and the even distribution of the top soil, especially if levelling has taken place recently or at some time. Water should also be allowed to run into an inspection hole to see how long it takes to drain away, in order to confirm that drainage is adequate. Further investigation should also be made to see if seepage from outside the greenhouse, or shed from roofs or higher land, is likely to be a problem. This is something which is frequently ignored, but which can give rise to considerable trouble, especially for early crops.

If the soil itself is of good physical quality, this gives a flying start although much can be done to improve soil in a limited area by adding peat.

CULTURAL METHOD SELECTION CHART

Condition of border soil and performance of previous crop (if any)	Condition of roots of previous crop	Heat and light levels	Control of environment and feeding	Decision on cultural methods
Soil good (new soil or sterilised). Previous crop performed well.	Good	Good	Good	Border culture: early or second early crop. Sterilisation not essential but advisable as safety measure for second crop.
Soil good (previous crop performance fairly good and pre-planting sterilisation carried out, or alternatively re-soiled.	Fair	Fair	Fair	Border culture for mid-season crop only. Sterilisation desirable.
Soil doubtful. Crop performance variable. Sterilisation or re-soiling impracticable.	Poor	Good	Fair	Avoid border culture. Ring, container (including peat mattress, polythene bags or trough) or straw bale culture advisable. Or alternatively use grafted plants.
Soil good. Previous crop performed reasonably well. Sterilisation impracticable.	Fair	Poor	Fair	Border culture for mid-season or late crop only. Grafted plants could be used but problems of growth in areas of poor light indicate container or straw bale culture for earlier crops.
Soil poor. Cropping results poor, or complete failure.	Very poor	Poor	Poor	Grow on ring culture, straw bale or in containers—cold or mildly heated crop only.

Presumably border culture will only be considered where the soil is new to tomatoes or has been sterilised (see page 200). Potato eelworm is the only really serious long-term malady likely to be a hazard to border cultivation in a 'new' soil. It should be realised that potato eelworm, affecting as it does solanaceous plants to which family tomatoes and potatoes belong, is likely to arise mainly from continuous cropping with potatoes, whether by the gardener or as a legacy from farming days. Eelworm can also readily be carried in on feet or by water, and can be blown in from neighbouring land. Soil analysis and eelworm counts are, I feel, very essential preliminaries to border cultivation.

Isolation of the greenhouse border soil from outside infestation, especially in respect of eelworm, is something which should be given serious consideration, although if the greenhouse foundation is of sufficient depth, this should suffice. More important for older borders will be the problems of deep-seated 'self-inflicted' troubles—root rots, potato eelworm and virus diseases from previous tomato crops—which are difficult or impossible to control completely even with sterilisation.

There are problems in installing barriers of either polythene or concrete, not least of which is drainage and, to be practical, it would be better to consider methods other than border culture if one had to go to these lengths.

Preparing the Soil

In theoretical terms all that is necessary is to provide the tomato with air, moisture and soluble nutrients and it will produce crops. I have seen excellent crops grown on this premise, both in Britain and on the Continent, especially in the Netherlands where tomatoes are grown on many of the reclaimed coastal areas in almost pure sand. Yet there is obviously great virtue in growing a crop in a soil of good structure, where air penetration, moisture movement and nutrient release are optimal compared with a 'played out' soil where the structure has been destroyed or deteriorated and the mineral particles, instead of existing in crumbs, are packed together. The fashion in which nutrients are released is also thought to contribute to the quality of the tomato fruit, and this indeed adds considerably to the quality of tomatoes grown in heavy northern soils, although slightly lower growing temperatures and a longer period of fruit formation cannot be ignored as one of the main reasons for this.

While, on a large scale, rotary cultivation or ploughing may be the desirable method of cultivating the soil, digging is much more likely to be the method favoured by the majority of smaller growers and gardeners. The reason for digging a soil at all is not perhaps fully understood, it often being carried out by sheer habit. The virtues of digging are best listed as:

1. loosening the soil and physically creating air and moisture channels
2. allowing the incorporation of bulky organic humus or conditioners
3. improving drainage by loosening the subsoil in deep digging
4. allowing easy planting in the loosened soil
5. inverting the soil, if digging is correctly executed, ie bringing up the lower area to the top and vice versa, which in theory should bring nutrient-containing soil to the area of early root development; *subsoil should be left where it is*
6. allowing weeds and surface debris to be put below the top spit of soil to act as a source of organic matter

Digging should be carried out during December or January, or before mid-November if chemical sterilisation is being carried out (see page 123). Whether deep or double digging is preferable to single digging is now perhaps the subject of some controversy. The roots of tomatoes are capable of growing to a great depth, as determined by underground observation windows at experimental stations. Yet deep roots remain in the soil and can result in the build-up of deep-seated troubles, especially virus diseases, despite sterilisation, which is generally only effective to a depth of 10–12in. The philosophy of digging only to shallow depths to discourage deep rooting (provided drainage is in no way inhibited) could have certain merits, especially as it will be seen later that tomato plants can produce excellent crops in relatively small quantities of growing medium.

During digging, any vegetative portions of obnoxious weeds should of course be removed, especially where couch or horsetail exists, as can often be the case in a 'new' soil. This raises the point that some control measures *outside* the greenhouse will also be necessary to prevent re-introduction. A word of warning here: do *not* use very soluble types of weedkiller such as sodium chlorate immediately outside a greenhouse, as this can readily seep through into the greenhouse soil. It would be much safer to use 'spot' weed control methods based on chlorthiamid (Prefix) or dichlobenil (Casoron G).

The Case for Applying Organic Materials

Farmyard manure (FYM). When applying any type of organic matter there are two basic considerations. Some types of organic matter contain fair quantities of nutrients (see page 60) while others do not. To apply FYM is to provide a source of nutrients and at the same time physically to condition the soil—physically in the sense that the sheer bulk of FYM separates the soil particles and creates channels for air and moisture movement. The conditioning is done by providing the raw material for humus production with all the benefits which this entails, crumb formation of the soil particles being of primary importance.

Peat, on the other hand, merely conditions the soil and does not supply any appreciable quantities of readily available nutrients (see page 80). Yet this can be a very great advantage, as one has all the benefits of conditioning without the complications of applying unknown quantities of nutrients, which after all can be much more accurately applied in fertiliser form.

A still further complication is that where FYM is applied an excessive amount of nitrogen may be made available to the plant, particularly if heat sterilisation has been practised, as there will be a rapid production of ammonia.

Well-rotted FYM should be applied at rates of up to 1 ton per 160–200sq yd or 1 cwt per 8–10sq yd. FYM can also contain weeds, pests and diseases, not to mention disinfectants. Wood shavings are often used in the bedding of animals these days, and if these are present in quantity they can give rise to some nutritional problems, such as temporary nitrogen shortage.

Municipal compost. Municipal compost is freely available in many districts and can be used at rates of up to one-sixth of the total bulk of soil. It can, however, contain quantities of heavy metals and so give rise to nutritional problems on a long-term basis.

Peat. Peat can be used copiously at the same rate as FYM (about 10lb or more per square yard), although it is frequently used less lavishly than this. It is difficult to reconcile weight with volume of peat, but in general terms and with peat of 60 per cent moisture content, there are 6–8 bushels of peat in 1cwt. Note that sphagnum peat is very acid and that where it is applied in quantity there will be a considerable lowering of the pH figure. To offset this, for every bushel of peat allow 5–6oz of ground limestone (5–6lb per cubic yard).

Other organic materials. The application of materials such as garden compost or brewery waste is also useful for soil conditioning purposes, although unless garden compost is properly made it can be very suspect. At the risk of offending organic gardening enthusiasts, let me hasten to add that *properly prepared* garden compost is a truly excellent material for applying to tomato borders and will result in fruit of a superlative quality. There are no upper limits to the quantities necessary, supply usually being the main problem. One word of warning, however, concerning the organic method of growing tomatoes: when pests or diseases are inadvertently introduced there may be a 'wait period' before natural balance is restored, and this could result in very serious crop loss. Gardeners anxious to know more about composting principles are advised to get in touch with the Soil Association.

Topping up the Water Reservoir and Flooding

After cultivation of the border, consideration must be given to the need for topping up the water reservoir and flooding to wash out excess salts, along with plant toxins, especially following steam sterilisation. There has already been discussion on this point in Chapter 7 and broadly speaking the following rules should be observed.

1. Where soil is new, or re-soiling has taken place, merely water to moisten the soil well 2 to 3 weeks before planting.
2. Where soil has been sterilised by heat, flood to wash out surplus ammonia 2 to 3 weeks before planting.
3. Where soil has been used for some years and is chemically sterilised, check salt content to ascertain whether flooding is necessary. If flooding is not necessary, merely water before planting.

When methods alternative to border culture are to be carried out, flooding does not have to be considered, the water is simply given to the growing plant as it needs it. It must be appreciated, however, that as fresh growing media are being used there should therefore be no soluble salt problem.

Application of Lime

Soil analysis will determine the pH figure and also give a lime require-ment. Ideally tomatoes in a soil-based medium like a pH figure of 6·5. Lime can be applied after any flooding, or lightly washed in near the end of the flooding period. There is a tendency these days to use magnesian limestone, which contains magnesium as well as calcium, but ground limestone can of course also be used. So, for that matter, can hydrated lime, but in amounts roughly equal to 75 per cent of the stated ground-limestone requirement following analysis. Where analysis is not carried out, it is necessary to guess how much lime is re-quired, and 8oz of ground lime per square yard is a normal application to soil, although it would be better to check the pH figure. Lime should be applied evenly, and if not lightly watered in, it should be forked in. The problem of contact between the lime and any FYM which has been applied, with the subsequent release of ammonia, always exists. If, however, the FYM has been well turned in, there is little immediate physical contact and therefore little danger. Nor in fact is there any real danger of ammonia release if either ground or magnesian limestone is used as neither of these materials are caustic, whereas hydrated lime is relatively so.

Ammonia release in gaseous form will of course damage any seeds or plants in the vicinity and also result in loss of nitrogen. There is also the further danger that if the ammonia is dissolved in the soil water it will be taken up by the plant, causing very rank growth if not more severe long- or short-term damage.

APPLICATION OF BASE NUTRIENTS

While fairly precise rules exist for the application of base nutrients to tomatoes, it should be pointed out again (see page 76) that it is often very difficult to make accurate allowances for the nutrient reserves in the soil. Many experiments have been carried out where no base applications at all were given to older tomato soils, the nutrient requirements of the plants being met as they grew. There is increasing favour for such procedure, even for *new* low-nutrient soils and media, applying liquid feed from the outset. It is, unfortunately, not possible to state which method is preferable or which will give greater success. Soil analysis, despite its failings, will at least give an indication of the existing nutrient level of the borders, and it will certainly provide an accurate soluble salt figure. Where salt contents are found by analysis to be of a high order and where flooding cannot be carried out, then obviously the application of most base dressings would be folly, as this would put the salt content at danger level. Conversely when soil analysis shows the soluble salt content to be *low* and nutrient levels fairly high, base dressings should be applied to bring the salt content to a level which will exercise growth control by evening up the osmotic pressures (see page 52). Nutrient applications (N, P, K) can, in this instance, be

Soil	Nutrient level as confirmed by analysis	Soluble salt	Advised course of action
Old tomato soil	Medium to high	Normal	Apply either no base at all, feeding the plants with liquid feed (at correct dilution) from outset or apply base dressing or magnesium to bring salt content to a pC of 2·8.
Old tomato soil	Medium to high	High	Try to flood if possible, but if not, **apply no base feed**. Liquid feed from outset according to appearance of plants.
'New' tomato soil	Low	Low	Apply base dressing at full rate, including magnesium sulphate.
Old tomato soil	Medium to high	Low	Apply base feed to adjust soluble salt content. Magnesium sulphate can also be applied.

waived, and magnesium sulphate only applied as the adjusting agent.

The various procedures are shown in the table on the previous page.

Types of Base Dressing

Tomato base feeds are available with 'high potash', 'medium potash' and 'high nitrogen' contents. High potash bases are applied following steam sterilisation, the extra potash being included to counteract the excess nitrogen produced after the sterilising procedure (see page 201). Medium potash base dressings are for soils which have either been sterilised chemically or not sterilised at all. High nitrogen base dressings are applied when nitrogen levels are likely to be low, eg, following a lettuce crop or when planting late. It will be noted that considerable importance is placed on the nitrogen/potash ratio, the phosphorus levels being more or less ignored.

Typical base feed dressings are given on page 64. The John Innes base is frequently used as a high potash base, but tends to be expensive. Early tomato growers frequently apply additional sulphate of potash and it is now standard practice to apply magnesium sulphate in addition for both early and late crops.

Quantities of Base Dressing to Apply

While soil analysis can be used as a practical guide, the more important considerations have already been outlined. The normal rate of base dressing can be reduced to 6oz per square yard when applying 3oz sulphate of potash. Magnesium sulphate is also applied at 3oz per square yard provided there is no salt problem.

For main crop tomatoes apply less potash (2oz), while for late tomatoes it is doubtful whether extra sulphate of potash should be applied at all, but the base fertiliser applications pushed up to 8oz on heat sterilised soils. Vigorous varieties may require high potash base feeds under all circumstances.

Base feeds, carefully measured out, must be applied evenly, as concentrations in any one area will give rise to salt problems which could be serious if plant positions happen to coincide with such areas.

A summary of the procedure is as follows:

Crop	Soil	Base type	Quantity per sq yd
Early and main crop tomatoes	Sterilised	High potash base	6oz
		Sulphate of potash	2–3oz
		*Magnesium sulphate	3oz
Early and main	Not sterilised or chemically sterilised	Medium potash base	8oz
		*Magnesium sulphate	3oz
Late tomatoes (either heated or cold grown)	Heat sterilised	High potash base	8oz
		*Magnesium sulphate	3oz
ditto	Unsterilised or chemically sterilised	Medium potash base	6oz
		*Magnesium sulphate	3oz

* Do not apply magnesium sulphate if there is a high soluble salt problem and flooding is not practicable. Modifications to these quantities may be needed for vigorous or weak varieties. Apply high nitrogen base feeds at 6oz per square yard following lettuce or other nitrogen-demanding crop.

Final Preparations

Normally the condition of the soil in greenhouse borders following cultivation and any flooding necessary is sufficiently good to allow base fertilisers to be applied before finally forking down the bed preparatory to planting. A light forking, followed perhaps by some consolidation to get rid of any lumps, and finally raking down to produce a reasonably level and fine tilth, should be carried out though it is unnecessary to go to extreme lengths.

GROWING GRAFTED TOMATOES—INITIAL PREPARATIONS

There is no difference in soil preparation, apart from the obvious fact that there is no need for sterilisation. There could be instances, however, where sterilisation is still worthwhile even when grafted plants are used, especially if the soil is very weedy; this matter is dealt with in Chapter 13.

PRE-PLANTING SUMMARY FOR BORDER CULTIVATION

1. An assessment of the previous crop performance and examination of rooting system to determine the best course of action for the next crop, including the need for sterilisation.

   Oct–Nov Chemically sterilise with methamsodium before 15 Nov (see Ch 13) for further details on sterilisation)

2. Soil analysis and eelworm count—paying particular attention to soluble salt level.

   November

3. Cultivation, including weed removal and subsoil loosening.

   December

4. Complementary to cultivation will be the incorporation of organic matter such as FYM, peat, etc.

   December

5. Flooding, if necessary or deemed advisable.

   Jan–March

6. Lime applied if necessary according to pH and lime requirement figure.

   Jan–March

7. Pre-planting watering if borders are too dry.

   Jan–March

8. Base dressings applied 7–10 days before planting.

   Feb–May

9. Forking bed and raking soil reasonably level.

   Feb–May

RING CULTURE—INITIAL PREPARATIONS

Ring culture (fig 18) is a method of growing tomatoes in growing medium in quantities which allow for complete renewal annually without great labour or expense. This conception of a two-zone rooting system, with the initial surface roots in the rings and, as the season progresses, a 'take-over' by roots which develop in the aggregate, allows a considerable measure of growth control difficult to achieve where roots are allowed to develop freely under the uninhibited conditions of border growing. These issues will become clearer as the cultural directions are discussed in more detail.

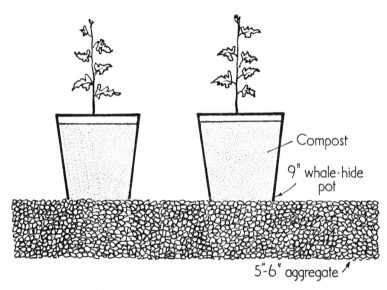

Compost

9" whale·hide pot

5"-6" aggregate

18   *Ring culture: 9in whalehide pots are set out on a layer of aggregate*

The 'clean' start that can be made with ring culture demands a hygienic approach from the outset. The existing border soil (if any exists in the first case) should be removed to a depth of 6–8in or more to make way for a 6in layer of aggregate, below which there should be ample provision for good drainage, and this could mean laying down some tile drains. As ring culture is usually adopted because of a history of crop failures, merely to put the aggregate on top of the remaining soil is obviously to leave the way open for further pest or disease problems, unless the existing soil can be isolated effectively. Potato eelworm is the main culprit, although there could also be virus diseases harboured in the root debris of earlier tomato crops. The more common fungal parasites are not likely to be such a problem, as they tend to exist largely in the top layer of soil. One therefore has to decide whether to lay down a floor of concrete or use a polythene barrier, in both instances ensuring that drainage is not inhibited.

Whether or not to chemically sterilise the lower level of soil is open to controversy. Subsoil of poor quality frequently lacks texture, making the dissipation of chemical soil sterilants difficult. Sterilisation should perhaps be confined to washing down the lower foundations of the greenhouse with formaldehyde (1 in 49). A deeper layer of aggregate (10–12in) may be an acceptable solution.

Selection of the Aggregate

There is a wide choice of material for the aggregate; the prime condition is that it should be an *inert yet porous* material. Well weathered ash, granite chips, coarse quality sand or gravel, or pebbles, can also be used. Ashes, however, frequently render up sulphur fumes which could be very damaging to the tomato plants and if used should be left outside in reasonably shallow layers for some time to leach thoroughly. Needless to say, there should be no contamination such as weedkiller in the aggregate chosen.

The aggregate should be put down in an even layer over the whole growing area; to leave uncovered patches of soil is to court trouble. It is, in fact, advisable to put the aggregate down over the whole floor area, and to lay conceete slabs on top of the aggregate to form a path.

Selection, Filling and Placement of Rings

The selection of a container for the growing medium is not a vital matter. The most usual choice is a 9in whalehide (bituminised paper) pot, usually without a base. Alternatively 9–10in clay pots with the drainage holes enlarged are frequently used, or rings may be made up of roofing felt or other suitable material, even linoleum. The 'rings' need not be round: boxes 10 × 10 × 10 inches can be made up specially and serve very well. The only proviso is that contaminated material must not be used.

The rings are spaced out at the decided distance apart, which will generally be in the order of 20–24in. They can then be filled to within 2in of the top with the appropriate compost, which is usually John Innes No 2. The recommendation that John Innes No 3 be used is frequently made, but the salt content of this is often far too high for young tomato plants and may cause root damage, apart from inhibiting growth. Soil-less media can also be used for ring culture, but liquid feeding must then be carried out from an early date.

Where rings or containers have bases, it is not necessary to fill them in position. It is frequently economically convenient to grow the plants in a warm propagating greenhouse for a period before setting them out in the growing greenhouse.

The quantities of compost required will vary according to the size of containers selected, but if 9in whalehide rings are used, approximately 12lb of John Innes compost will be required to fill them to within 2in of the top.

Note that the growing medium both in the rings and the aggregate should be sufficiently warm (56° F or 13° C) to receive the plants when they are set out.

Summary of Preparations for Ring Culture

1. Remove border soil and make adequate provision for 'isolation' from any diseased soil.
2. Install inert aggregate 6–8in deep (deeper with doubtful subsoil) and set out rings on top of this, filling either *in situ* or elsewhere.
3. Allow suitable time for both the growing medium in the rings and the aggregate to warm up (to 56° F, 13° C) before planting.

Special Aspects of Ring Culture

In the culture of tomatoes in borders, the plant has an uninhibited root run from the outset. In ring culture (and in some other systems) the roots are restricted in their early growth and this tends to act as a brake on vegetative development of stem and leaf. The vital difference between ring culture and other methods of culture involving limited quantities of growing media is that while some systems contain the roots in the medium throughout the season, strong roots run into the aggregate with ring culture, this usually coinciding with the period when the first truss of fruit has set and the swelling fruit acts as a natural brake on rank vegetative development. The root system which forms in the aggregate can in time take over the full growing function, but this transition must be allowed to take place naturally.

GROWING IN LIMITED QUANTITIES OF MEDIA

There are a considerable number of methods under this system and each is best dealt with individually, highlighting similarities with other methods.

Polythene Bag Culture

This involves the use of 12in diameter black polythene bags, 250-gauge, with drainage holes, which hold approximately half a bushel of growing medium. The bags can be spaced out after filling at normal planting distances or filled in position, being placed either directly on the soil or preferably on a layer of polythene. Alternatively they can be placed on a layer of free draining material such as inert gravel or polystyrene through which a drainage tile runs (fig 19).

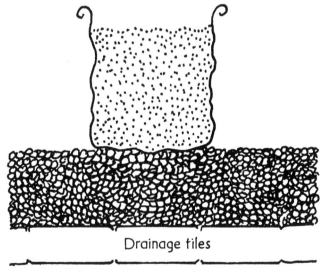

Drainage tiles

19 *Polythene bags are placed on a layer of free-draining material or a layer of polythene. Variations of this method include the use of 'bolsters'*

There are variations of this method, including the use of 'bolsters' where bags of compost are laid in rows and upper sections cut out to

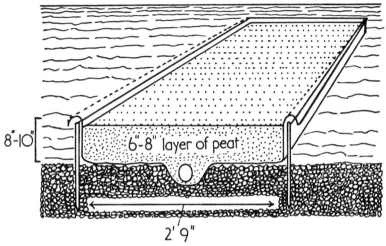

8"-10"

6-8" layer of peat

2' 9"

20   *Make-up of polythene trough to take two rows of plants*

allow insertion of plants. Drainage is provided by making slits in the sides of the bags. Bag or bolster culture lends itself to the use of soil-less media, but unless strict attention is paid to detail there can be nutritional problems, especially the build-up of soluble salts (see page 197).

Growing in Boxes of Soil

This is a method favoured for smaller-scale growing by amateurs, whereby boxes of a suitable size are filled with the selected growing medium, generally soil-based. Boxes should preferably be at least 12–14in in depth and of similar width and have drainage facilities in their base. These should be treated with spirit-based preservative or Woolmans Salts (1lb copper sulphate in 1 gallon water), but definitely *not* with creosote, and there should be provision for free drainage, usually achieved by elevating the boxes on bricks. Apple or orange boxes are frequently used on a short-term basis, and these suffice admirably.

Trough Culture

This system involves the making up of polythene troughs filled with peat or peat/sand media, with or without drainage tiles running through them (see fig 20). Various widths of trough can be used. If there are no drainage tiles slits are made in the sides of the trough.

Another method involves the use of troughs made of expanded metal, large tiles or wood, and supported on blocks so that surplus moisture can drain away readily. Soil-less media are usually used and culture is largely similar to polythene trough or bag culture.

Peat Mattress Culture

Here 9in whalehide pots filled with growing media are spaced out on a layer of polythene, 2ft to 2ft 6in wide, with upturned edges (using strained wire and pegs) on top of a 2–3in layer of peat to which has been added 6–7oz of ground limestone per bushel plus 1½–2oz of general fertiliser per bushel. Levington compost has been used with some success in the whalehide pots for this system. The plants contain their roots initially in the whalehide pots and feeding is applied to these; when roots develop in the peat, feeding should be applied to the mattress. Variations of this system have been practised for many years.

Growing in Whalehide Pots on Top of Soil

Where short season crops only are concerned, plants can be grown with reasonable success in 9in whalehide pots with bases, filled with growing medium, and placed on top of the greenhouse border. While any pests and diseases in the border may in time affect the plants, with a *short-term crop* this is not usually of great consequence. This is not a method, how-ever, which can be recommended with any great degree of confidence, and it would be better to lay down polythene and set the plants on this, mulching with peat, which will result in a modified peat mattress culture system.

Growing in Polythene Buckets

Soil-less media, especially peat and vermiculite, has proved most successful in polythene buckets (2 gallon size), provided with three 1in drainage holes. The buckets are best placed on some free-draining aggregate such as vermiculite, gravel or coarse sand. In essence this is a form of ring or polythene bag culture, but unlike ring culture there is only a very limited root development into the aggregate initially.

Growing in Layers of Growing Media on Top of Polythene or in Polythene Trenches

Isolating the border soil with a layer or trench of polythene, a 6–8in layer of growing medium is spread on top of or into this. Soil-less media are best for this system, but the quantities required are rather large. Drainage in trenches will require careful consideration, and could be provided for with plastic draining pipes. This is a system which has proved very successful in Ireland.

STRAW BALE CULTURE

Where the use of border soil is impracticable due to deep-seated virus infection or eelworm infestation and other methods are not acceptable, straw bale culture is worth considering. There are tremendous advantages in using straw bales, not least of which is the heat generated by the decomposing straw and the CO_2 enrichment. It is also possible to have a quick turn round from one crop, such as lettuce, to tomatoes, without recourse to sterilisation (growing in containers or trenches of course also allows this).

The harder-textured bales of wheat straw are preferable, as they retain shape better, failing which either oat or barley straw will suffice. It is better if the bales are wire bound as string rots, permitting early collapse. While ideally the straw should be fresh, it matters little if it is slightly weathered but bales stacked outdoors preparatory to use should be protected by polythene. The weight of bales varies, the average being 50–60lb. Straw treated with TBA or Picloram is unsuitable.

Wads of straw 7–10in thick, can be used in preference to whole bales, and this method is of special interest in amateur growing where height is frequently restricted.

'Composting' the Straw

Bales or wads of straw are placed in the most suitable position, taking into account the size of greenhouse involved. Commercially either single or double lines of bales, end to end, are placed in rows 5ft apart, but this is frequently unsuitable in smaller greenhouses. My own greenhouse is 15ft wide and I find that two single lines of bales on each side and a double line in the centre is a convenient method of growing (see fig 21). This may allow fewer plants in the greenhouse (compared with other methods), but to compensate for fewer rows the plants are spaced closer, on average 3 per bale (or 1 per wad). The bales or wads are placed on layers of polythene (old polythene sacks are useful) to prevent contamination from greenhouse borders. Bales are laid flat to avoid toppling and it helps to conserve both space and water if they can be placed in a shallow 1½–2in depression. Take care to keep the wire binding joints upwards to avoid puncturing the polythene.

About three weeks before the planting date, sufficient water is applied to thoroughly soak the bales, it being found that this is better carried out on several occasions over the space of a few days, and about 10 gallons of water per bale will be required—and *pro rata* for wads. The vents in the greenhouse are now closed and, if the weather is cold, the heating system is put into operation to provide about 50° F (10° C).

Fermentation Process

Considerable modification in treatment has taken place in recent years, the main difference being that smaller quantities of fertiliser and a shorter period are required for the fermentation process. Working out the quantities of fertiliser needed to induce fermentation can be difficult, owing to the varying weight of the bales. Allowing the average figure of 50–60lb per bale, for each bale (or again *pro rata* per wad) apply:

> 1lb ammonium nitrate-lime (nitro-chalk or similar)
> 6oz triple superphosphates
> 6oz magnesium sulphate
> 12oz potassium nitrate
> 3oz ferrous sulphate

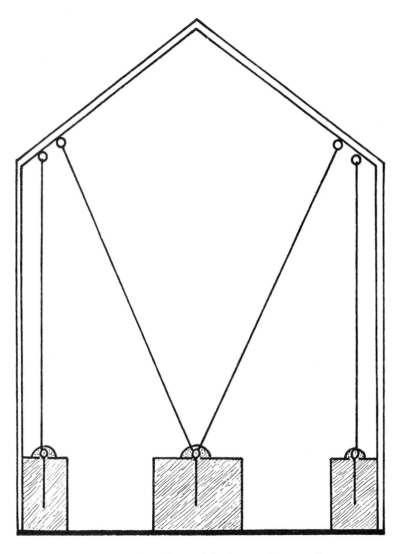

21 *Straw bale culture showing bale placement down greenhouse*
and a suggested training method

I

Earlier recommendations were for a longer decomposition period. Here ammonium nitrate-lime (Nitro-chalk, Nitram, etc) is applied at 1½lb per bale after the straw is wet, followed by 1lb in 3–4 days, followed in another 3–4 days by the fertiliser quantities previously listed—but reducing the ammonium nitrate-lime to ¾lb. Alternatively 1½lb of any compound fertiliser can be used for the final dressing with excellent results, in addition to the ¾lb ammonium nitrate-lime.

A small ridge of soil-less medium—3 parts peat, 1 part sand—is run along the bales or wads after the final application of nutrients. This ridge merely needs to be large enough to accommodate the plant's root ball, and need not contain large amounts of fertiliser but requires lime, about 4–6oz per bushel. A temperature check should be made on the bales and it will be found that the centres will reach 110–130° F (43–54° C). When the temperature falls to around 100° F (38° C) planting can take place.

Planting Procedure

Plants are set out at 12–14in apart (3 per bale) and angled outwards, if practicable, so that they can be trained V-fashion. Wires can be used for support as the bales rot and drop, but if string is used this should be tied very loosely. Further implications of straw bale culture are discussed in Chapter 11.

SINGLE AND DOUBLE TRUSS CROPPING

There is current experimental interest in the cropping of tomatoes on a single and double truss basis, or on a 3 crops per year cycle. Various equipment is used for this technique, tiered troughs being the most popular, these being fed and watered semi-automatically through polythene pipes which serve for watering, feeding and sterilising. Only the first or second trusses are allowed to form, the plants being pinched out beyond this point.

The theory of this technique is quite simple and centres around the premise that a square area of cropping in a greenhouse is capable of producing only so many pounds per year on conventional cropping, say 3lb per square foot, taking the plant to 10 trusses or more. It should be possible to exceed this by having very closely set plants yielding $\frac{3}{4}$–1lb each. With three crops per year, there is scope for tremendous yields.

Gardeners wishing to experiment on their own should set out some plants in the border at 10–14in apart each way, and by supporting them with canes or netting, allow two trusses to form. Poor light, however, may limit the production to two crops per year instead of the intended three, and there are soil sterilisation problems to consider.

9

Planting Procedures

Cultural procedure follows a remarkably similar pattern no matter what method of culture is adopted and here I propose to deal with cultural techniques on a broad basis, making suitable comment on differing methods at the appropriate point. Details of modification follow at the end of the chapter.

PLANTING DISTANCES

The optimum number of plants per given area is a matter of considerable controversy. Some authorities recommend very close planting, whereas others maintain that yields per plant are greater when crops are less dense and each plant receives a better quota of light. Three factors must be considered when discussing planting distances: (1) the district, as I feel certain that areas of better light are more suitable for dense planting; (2) the variety, because healthy vigorous varieties take up more room than less vigorous ones; (3) the type of greenhouse and its orientation, a factor that does not affect density but has a considerable influence on placement. Where the greenhouse is east-west orientated a too-close longitudinal planting along the south-facing side will successively shade each row of plants and it is therefore better to have the plants more widely spaced longitudinally and closer on the east-west axis. Conversely north-south orientated houses are better planted more closely on the north-south axis and less closely on the east-west (see fig 22). The object in both cases is to allow better entry of solar radiation, the main source of which is of course from a southerly direction.

Variation in plant densities is generally practised commercially rather

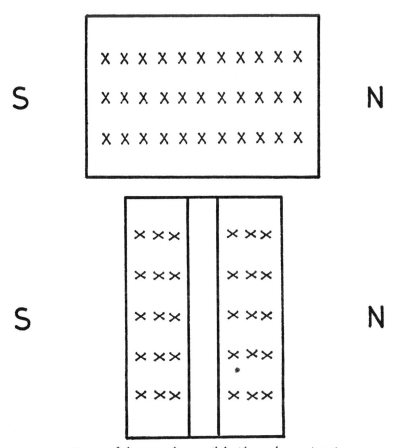

22 *Spacing of plants must be reconciled with greenhouse orientation*

than in amateur circles yet with the better all-round light admittance of the smaller amateur greenhouse, the reverse could well be the case if it were not for the tremendous wastage of space occasioned by paths in proportion to total area. To appreciate this, think in terms of the total area of the greenhouse, eg 8ft wide by 12ft long, which is 96sq ft. A path is necessary down the length of the greenhouse, and this is often as much as 3ft wide: 3 × 12ft = 36sq ft, which is over one-third of the total ground area. In commercial units it is usual to have path areas only about one-quarter to one-fifth of the total ground area (ie leaving 75–80 per cent for cropping).

Commercial densities allow for around 3·3sq ft per plant, including paths, but on the premise that there must be a greater allowance for paths in smaller amateur houses, about 4sq ft per plant there is reasonable. To estimate the number of plants per greenhouse, take the total area and divide by 4. If one were planting square this would mean 2 × 2ft per plant, but it is generally better to alter this according to the orientation pattern previously referred to, running the plants out to 2½ft apart longitudinally in an east-west greenhouse, and 14–18in apart across the house (see again fig 22). Planting distances are however not absolutely vital and it matters little whether the plants are a few inches closer one way or another.

Excessively dense planting can bring problems of management and encourage disease by restricting air movement. 'Compact' types of tomatoes can certainly be grown more closely than vigorous types.

Spacing Modifications According to Cultural Methods

On straw bales, as we have seen, there must obviously be a closer planting distance, congestion in the rows being overcome by modification of the training system.

Where trough culture systems are practised it will invariably be necessary to space the plants more closely in the rows, irrespective of greenhouse orientation, although it is, of course, sometimes possible to plant across the greenhouse. Double rows are usual in troughs.

Soil Warmth and Plant Spacing

An issue of some importance with regard to plant spacing is the need to ensure that plants are set out in warm soil or growing medium—not less than 56–57° F (14° C) at 5–6in depth. Where a heating system is specifically designed to give warmth in the immediate vicinity of plants this temperature can be reasonably easily achieved, whereas with perimeter or free-discharge warm air systems it may take a considerable time for suitable temperatures to be reached. The need to avoid growing checks to young plants, newly set out, cannot be over-emphasised, as this will invariably result in early infection from parasitic diseases, with serious long-term consequences (see page 177). It is for this reason that small-bore pipe systems have become so popular in commercial spheres, as they allow the heat to be taken to the plants. On the Continent much use is made of underground alkathene pipes installed appropriately to give the necessary soil warmth without the need to 'force' the heating system (fig 23). Electric soil-warming cables, however, are much more practical on a smaller scale.

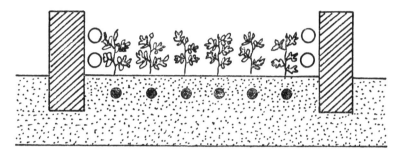

23   *Underground alkethene pipes for soil warming are widely used on the Continent, but less so in Britain*

Soil warmth does not cancel out the need for suitable air temperatures, and this can only be accomplished with a well-designed heating system (see pages 33–34 and Chapter 3).

THE RIGHT TIME TO PLANT TOMATOES

Recourse to the timetable in Chapter 6 will give information on the normal programming for the crop. But there must be a critical appraisal of the *actual* compared to the theoretical temperature and light levels in the greenhouse in question. I am frequently asked to advise on the production of early tomatoes when reference to some literature has elicited the fact that tomatoes are planted in late January and early February in favoured climatic areas of good light. But to follow such procedure in a badly designed greenhouse in a poor light area would be folly, even if the necessary temperature levels could be readily enough achieved.

There is much virtue in a case like this in paying heed to local trends. Obviously if there are a number of tomato growers in the area able to plant in late January or early February and obtain ripe fruit in late March or early April, then there are grounds for thinking that such practice can fairly readily be emulated. Some districts fall into the category of 'late' areas because of poor light occasioned by latitude, industrial pollution, or sun shut-off by hills, and it would be very difficult to try and grow a really early crop of tomatoes under these conditions without recourse to artificial light. Good light intensities are essential not only to maintain photosynthesis at an acceptable level, but to encourage the formation of viable pollen. As the days get longer and lighter it is obviously easier to achieve this. A sensible appreciation of both natural light levels and the level of heating possible in the greenhouse and, more important still perhaps, the design of the heating system, should allow a planting time to be decided with fair accuracy. Mid-February to mid-March is most usual commercially, mid-March to mid-April for amateurs, and May for cold-grown crops.

At all events, and at the risk of repetition, *it is essential to pre-warm either borders or the medium in which the plants are to be grown,* and this could take 10 to 14 days or more, especially early in the year where the heating system is merely of perimeter design, or 4 to 5 days where there are soil-warming facilities. Unheated crops should not be planted until suitable soil temperatures have been achieved by solar heat.

Temperature Levels in Greenhouse

Reference to the table on page 97 gives the desired temperature levels for the requisite stage of growth. It is unlikely that precise temperature levels of the order quoted will be constantly achieved by amateur gardeners, and in general terms around 60° F (16° C) day and night is a reasonable target, especially for northern growers. Solar radiation generally raises the daytime temperatures, whereas at night the temperature usually drops. In addition there are some critical aspects of temperature control which are worth commenting upon.

Extremely high daytime temperatures lower fruit quality, especially as ripening commences. This is thought to be due to the drawing of moisture from the fruit to cope with the excessive temperature and the dry atmosphere frequently associated with it. Warm air is, however, able to 'hold' more moisture in gaseous form than colder air, and where an especially warm day is followed by a cold night, the gaseous moisture condenses on the cold glass and on the plants, causing a rise in humidity. Obviously, therefore, there is considerable virtue in giving sufficient heat at night to avoid condensation and its attendant high humidity. Excess temperatures are difficult to cope with in physiological terms. Plants, like animals, adapt themselves to a particular tempo of growth. Alter this suddenly and adjustments must take place, but some form of debility also occurs. More will be said about this in Chapter 12.

Stage of Plants for Planting

Ideally, plants set out from late January to early March should have the first truss of flowers fully developed and one or two of the flowers fully opened. If setting occurs on this truss soon after planting, this acts as an excellent brake on rank vegetative development. For later planting it is also desirable, but not so essential, that the first truss should be showing, as, with better light, growth is usually more restrained and balanced. Nevertheless there is still virtue in this procedure, especially when soil has been steam sterilised and there is likely to be a flush of nitrogen. Apart from the restriction in vegetative vigour, occasioned by a swelling first truss, there would appear to be no physiological reason why plants are not capable of survival and good development if planted at various stages, although ideally a sturdy plant which stands about 9–12in high (including pot) is desirable.

PLANTING PROCEDURE IN BORDERS

Marking out preparatory to planting should be done with some precision, using either a steel tape or a measuring board. A garden line can then be used to keep plants in line, the actual placement of the plants being marked either with a cane or, less accurately, by indenting the soil. A frequent practice is to paint the heating pipes with white marks at the appropriate point. Alternatively plants can be placed on the soil at their approximate positions, taking out holes, accurately spaced, and leaving these open for a few days before actually planting. This allows the soil to dry out a little if too wet, and at the same time the soil in the sides and base of the hole is warmed by solar radiation and air currents. This is important as these are the areas which will be first to contact the roots of the tomato plant.

Planting holes will vary in size according to the size of pot or block and can be taken out either with a trowel or a planting tool. Planting depth should be such that the seed leaves are an inch or so *above* soil level. Plants are watered before planting, and where plastic or clay pots have been used the root ball is carefully knocked out of the pot by tapping upside down on a convenient hard surface. Paper or peat pots are better not removed, and blocks are of course planted intact.

NOTE THAT ACTUAL PLANTING SHOULD NOT COMMENCE UNTIL THE SOIL TEMPERATURE AT 4–6IN IS 56–57° F (14° C) AS REGISTERED ON A SOIL THERMOMETER.

Shallow planting should be avoided as it allows the root ball to dry out; deep planting, on the other hand, can have its merits provided the soil is warm, and especially when, due to delays, plants have become very leggy. Layering can in this case be practised, setting the plant on its side and running the plant stem along a trench removed with a trowel, the object of this being to keep the bottom truss at a reasonable level and at the same time encourage adventitious rooting along the stem length. After setting the plants in position they are firmed up with reasonable pressure.

Points to Note at Planting

1. Soil at planting time should be neither too wet nor too dry, but merely moist.
2. If soil is too wet planting should be delayed and the heating kept in full operation to dry it out.
3. If the soil is too dry apply a light overhead spray of water.
4. Soil should be neither too firm nor too loose, merely moderately compressed.

After planting, give plants a light watering in, about ¼ pint per plant with plain water, or slightly more if the soil is very dry.

PLANTING SUMMARY FOR BORDERS

Planting distance: related in part to district; closer in good light areas, wider apart in poor light areas. Plants arranged, according to greenhouse orientation and cultural method, to avoid excessive shading to south. Average planting allowance for small greenhouses of 4sq ft as 2 × 2ft square planting, or 14–18in apart in rows of 2½–3ft apart.

Planting time: related to district, light and heat levels. On average from mid-February to early March in good light areas; March to April for amateurs with heat; May for cold crops.

Soil temperature: essential for it to be around 56–57° F (14° C) at 4–6in depth, necessitating well-designed heating system or a suitable 'wait' period—especially for unheated crops.

Temperature levels: see page 97 for precise temperatures, but on average 60° F (16° C) day and night is reasonable in amateur greenhouses.

Planting: with a trowel or planter; seed leaves 1in or so above ground level, unless plants are very leggy, when they can be layered. Plants should be moderately firmed.

Watering in: ¼ pint per plant, or more if soil is dry.

MODIFICATIONS FOR DIFFERENT CULTURAL SYSTEMS

Planting procedure is basically similar whatever system of culture is practised.

Ring Culture

Containers are placed in position, filled with compost which is allowed to reach the requisite 56–57° F (14° C) before the plants are set out in the centre of the rings in the same way as for border culture, the plants being watered in similar fashion.

Straw Bales

The ridge of growing medium along the top of the bales or wads (see page 134) will usually achieve the necessary temperature without any difficulty. Plants are normally set out at 3 per bale, with the root ball covered, and *well watered in.*

Containers and Troughs

Whatever type of container is involved, there is little need to vary planting procedure. The same is true of troughs. Care should be taken with soil-less peat-based media to avoid saturation of the compost and thus inhibit rooting by excluding air. The 'feel' of peat is the best guide to watering needs (squeeze a handful tightly to see if it retains its shape). Some peats dry out badly and are difficult to re-wet, leaving plants sitting high and dry.

10

Establishment and Training Procedures

EARLY TREATMENT OF PLANTS

Whatever system of culture is used, it is fundamental that the plants become established. By this is meant the development of roots from the root ball, which exists in the pot or soil block, into the growing medium. The physiological processes involved in rooting are neither easy to understand nor to explain in detail. There must be a balance between the leaves and the root of the plant, and there is a close relationship between them. Allow the leaves to transpire to excess, as they will do in a hot dry atmosphere, and this will slow up the formation of new roots. If the root ball is kept too dry this inhibits root development and simultaneously restricts leaf development. Other issues such as the salt concentration of the soil water will also influence the rate at which roots develop, and should the salt concentration be too high, the growing point of the roots will fail to extend, and this is also true of root hairs, the production of which, from existing roots, could be seriously affected. Excessively high salt concentrations in the soil can cause damage to roots by actual burning.

If the soil temperature should be too low, or the soil is too wet, roots may fail to develop, irrespective of all other considerations. Die-back of the roots can in fact take place, allowing the entry of parasitic fungi. It is perhaps difficult to understand why, in a sterilised (or, more accurately, pasteurised) growing medium, these parasites should exist in the first case, until it is appreciated that they are dispersed freely in the atmosphere and quickly invade all growing media, seeking a receptive host. Checked roots come into this category (see also Chapter 12).

Conditions for Establishment

Taking all these facts into account, it is a relatively straightforward matter to understand what conditions are necessary for ideal establishment. They can be summarised as follows:

1. High humidity of the atmosphere to reduce transpiration. This is achieved by restricting ventilation for several days and frequently damping-down. Automatic ventilation systems or fans should therefore have their operating thermostats adjusted accordingly.
2. Warm soil or growing medium to encourage root development. Warm air to encourage leaves.
3. Just sufficient water to sustain growth with no wilting and without over-watering, so excluding air.
4. Salt solution of soil at the correct level so that plants are able to assimilate the right amount of water and nutrients. This can be ensured by accurate feeding, by making sure that liquid fertilisers are used at the correct dilution, and by avoiding excess of solid fertilisers (see below).

The basic philosophy underlying the various treatments suggested above is designed to encourage plants to develop a root system capable of sustaining the rapid vegetative growth made by the foliage. Obviously, however, if there is no restriction on the root development, vegetative growth often tends to be lanky and unproductive. Dry regimes, as practised by some, endeavour, by limiting the amount of water available, to limit vegetative growth accordingly. It is a technique practised successfully by many, and unsuccessfully by many others. Some dividing line must be drawn between the restriction of roots and enough early development to sustain growth, and it is here that containers come into their own, as the early root formation will be limited to the boundaries of the containers, and this in itself tends to have a braking effect on foliar development and encourages the plant to produce flower trusses, which it may not do if too rank.

It is usual to give only sufficient water to prevent wilting until it is seen by freshening of the growing point that the plant has 'taken' and is starting to draw on the available nutrients in the soil. Different systems of watering will affect the rate of root development, and certainly gross watering will encourage surface and discourage basal rooting to the long-term detriment of the plant as the season progresses. Trickle irrigation systems tend to produce a cylindrical type of root, especially on light free-draining soils. Soil-less media, especially if all peat, produce 'feather' roots, different in character to roots produced in soil.

Osmotic Feeding

It is at this stage that consideration can be given to the technique of osmotic feeding. To go back to soil preparation, it will be remembered that reference was made to the adjustment of soluble salt levels to pC 2·8, so that there would be a measure of osmotic equilibrium, which in itself would tend to restrict growth even where quantities of nutrient and supplies of water in excess of the plants' needs are in reserve or given. To carry this a stage further, liquid feeds accurately diluted (pC 2·8 to 3) will help to sustain this osmotic equilibrium. The situation is still further enhanced if liquid feeds are balanced to the needs of the plant in respect of potash and nitrogen levels (see page 159). In general terms, high potash feeds are usually required early in the year, particularly when a heat-sterilised soil-containing medium is used.

'Weaning' the Plants

After the initial period of high humidity and establishment, the temperature should be kept at the appropriate level, and water more freely given according to weather patterns. Obviously the precision of temperature control will vary considerably, and where there are no automatic aids to ventilation and heating, it is simply a question of attempting to achieve as much regularity in temperature as possible.

Reminder of temperature levels overleaf.

For Southern Growers

Stage	Night		Positive day		Ventilation commences at	
	°F	°C	°F	°C	°F	°C
1	58	14	64	18	74	23
2	60	16	64	18	74	23
3	62	17	65	18	75	24
4	62	17	64	18	70	21

For Northern Growers

1	56	13	70	21	74	23
2	56	13	68	20	74	23
3	56	13	68	20	74	23
4	60	16	68	20	70	21

Marginal variations are not highly important, and indeed must be accepted in many instances. It would be difficult to state categorically that unless temperature levels are precise the crop will suffer in one way or another, as no one grower or research worker can state the ideal temperature pattern for any particular area, although he may have established that good results will invariably ensue if the recommendations given above are reasonably adhered to. It is certainly the case, however, that where massive temperature variations do occur, as will certainly be the case in a completely unheated greenhouse, there can be undesirable side effects of which disease is obviously one. There will, as stated earlier, be physiological problems when a plant is expected to cope with immense temperature variations, and these are concerned not only with the processes of respiration, transpiration and photosynthesis, but with all related processes, especially nutrient uptake and the utilisation at night of carbohydrate manufactured during the day. The effect of temperature variations may be short- or long-term, and more probably a combination of both, but it is fairly certain that the most

Page 149 (*Above*): A good bottom truss is produced on plants raised at lower propagation temperatures. Note also the 'ghost spotting' of fruit due to moisture condensation. (*Left*): The effect of higher propagating temperatures—small but early lower fruit trusses

Page 150   (*Above*): Tomato plants newly set out on straw bales. (*Below*): A well-developed adventitious root system is typical of plants grown on the straw bale system

vital function of all, which is the production of flowers followed by successful fertilisation, will suffer in some way.

SOME ASPECTS OF WATER APPLICATION

The earlier waterings will, as we have agreed, be more or less confined to direct application in the vicinity of the root ball. Confusion frequently exists concerning the so-called searching of plants for moisture. Plants may not be capable of logical thought (although some research scientists think otherwise!) but they do respond to certain influences. The rate at which plants extend their roots from the root ball into the border or growing medium and subsequently develop them, is governed largely by the quantities of water and nutrient present in the vicinity where the plant roots are actually growing, and not merely because they seek it out. Perhaps there is rather a thin line of demarcation between the two principles, but in practical terms roots will actually develop better in areas of adequate moisture and nutrient.

ESTABLISHMENT-PERIOD WATERING PROCEDURES FOR DIFFERENT CULTURAL SYSTEMS

Cultural system	First waterings of plants	Moisture state of growing medium	Humidity of atmosphere
Border, including grafted plants	Minimal watering in area of root ball	Keep merely moist by damping	Frequent damping with fine rose overhead, especially in mid-morning
Ring	Adequate watering of root ball frequently	*Light* waterings to be given in addition to whole growing medium, which may dry out otherwise	*Very frequent* damping necessary due to the dry atmosphere which usually prevails
Straw bale	The whole of the growing medium	Keep bales wet by use of hose or	Wetting the bales usually helps to

Cultural system	First waterings of plants	Moisture state of growing medium	Humidity of atmosphere
	ridge is usually kept moist when bales are watered	overhead sprays	keep the atmosphere sufficiently moist, but damp down also if necessary
Polythene bag, trough and trench culture	Only minimal watering required, in area of root ball	Peat based media usually retain their moisture content well, provided they are sufficiently damp at planting time	Damp down as for border culture
Peat mattress	A compromise of treatment between ring and bag culture. *Light* ball waterings only	As above, but ensuring that the small amount of growing medium in ring does not dry out	If peat mattress is moist this usually ensures a sufficiently high humidity but damp down on bright days
Container systems	Where soil-based medium is used, treat as for border culture. For soil-less medium exercise the same care as for bags (above)	Large bulks of growing medium will usually require no further watering, apart from damping, until plants are established	Frequent damping down, especially for raised box systems

Note: the subject of watering and nutrition, both early and late in the season, is dealt with in detail in the next chapter.

SUPPORTING AND TRAINING

Tomato plants in their natural habitat will, as stated earlier, sprawl about on the ground and have a bushy many-branched habit of growth.

Cultural systems under glass demand a rigid system of control which involves support, shoot removal, and eventually a form of pruning or leaf removal.

Support

Plants require support from an early stage, and while I am unaware of any experimental work concerned with the actual effect of early or late support, the fact remains that once plants are supported they appear to grow faster. Plants are better supported within a day or so of planting, although it does not appear to do much harm if support is delayed until the odd plant topples, which it will tend to do when heavy damping down has taken place. There are various ways of supporting plants; a popular method in amateur circles is to tie the plants to tall canes, but this system obviously has many drawbacks, especially on hygienic grounds. While methods of supporting plants can vary considerably, the use of strained horizontal wires at a height of around 6–8ft from the ground is basic, and the position of these should coincide with the rows of plants, unless the V-training system employed in straw bale culture is adopted, when they must be positioned accordingly after check measurement, from a point 12–20in from the ground level, to overhead wires in both directions (see again fig 21).

For support a 3- or 5-ply fillis or Italian hemp is used, although recently polypropylene twine has become very popular, as it is extremely durable. It is highly important that any wire used overhead must be of sufficiently thick gauge and secured firmly to the greenhouse. Some modification to the structure of the greenhouse is often necessary to ensure this. The fillis or string should be tied in a loose loop with a non-slip knot above the cotyledon (seed leaves) but below the lower leaves.

Plant hook wires 3/16in thick and 18in long are advisable for straw bales and lateral training systems generally, in the first case to avoid strain on the plant roots imposed by the decomposing bales, and in the latter case to keep the first truss of fruit from trailing on the growing medium. Alternatively a horizontal wire can be run along, about 12in above the growing medium, supported by stout wooden pegs, and this is often more practical for oblique and layering systems.

Plant Training

Plants are usually, but not necessarily, twisted clockwise round the string, taking care not to snap off the growing point, something which can easily happen when plants are turgid or growth is very hard (see page 159). There are several methods of training plants, best explained diagramatically (see fig 24).

Vertical training (A). The plants are allowed to reach the horizontal wires before being stopped. This system is frequently used in amateur circles on any system of short-term culture. Where it is carried out on longer-term crops, the plants can be arched over the wires (A2).

V-training (B). This system, where the plants are trained out obliquely and alternately in different directions, is useful with straw bales as the plants are set so closely together, although vertical training can also be carried out.

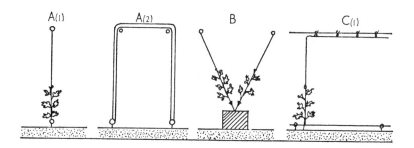

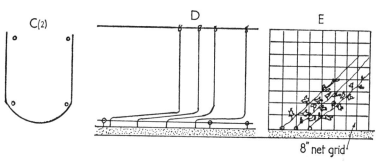

24   *Plant training methods*

S-hook system (C). This system is becoming more popular and involves the use of 16-gauge S-hooks set 14in apart, the plants being supported so many hooks along and dropped a hook as the season progresses, so that the plants are progressively layered. Wires or plant hooks are desirable to avoid the bottom truss trailing on the growing medium. At the ends of the rows the plants are turned on to the next row (C2).

Layering (D). Here the plants are initially trained vertically, being completely detached when they reach the horizontal wire, and layered— usually at the stage where lower trusses are picked and where the ripening trusses are kept well above growing-medium level.

Lateral training (E). From the outset plants are trained obliquely at an angle of 35–40°, wide mesh or polypropylene netting being a useful supporting medium for accomplishing this without a great deal of string manipulation. The plants are tied on to the netting with small loops of string or wire/paper clips. It is usual to lower the plants as the season progresses. This is however easier if slip lateral strings are looped round vertical strings. A new system of lateral training developed in Lancashire using interlocking wire hooks has proved very successful, as it readily allows progressive layering.

Merits of Different Training Systems

The choice of any particular training system will depend not only on the length of the growing season, but also on the type of greenhouse and whether ventilation can be effectively carried out to overcome the problem of disease which occurs more readily in systems other than the vertical. Laterally trained plants from the outset appear to be most prone to disease attack, due to air movement restriction, and plants layered later in the season eventually suffer a similar disadvantage.

PLANT PRUNING

Side-shoot Removal

This must be carried out from the outset, restricting the plants to one main stem, unless under special circumstances such as filling up a gap left by a plant loss. Removal is best effected when the side shoots are young, and preferably when turgid (full of moisture). There is little problem in the whole process of side-shoot removal, the little shoots merely being snapped off with the finger and thumb. Some varieties tend to form two leading shoots and it is necessary to decide which one is to be removed. This is not a vital matter, although if only one shoot bears a flower truss, let this one remain. Suckers forming at ground level should be removed, this especially so with grafted plants. There are various theories put forward that side-shooting with the fingers avoids the spread of virus disease compared with the use of a knife, but I find it difficult to accept this. Indeed the use of a knife dipped regularly in a mild sterilant could be more desirable than virus-infected fingers.

When shoots are allowed to get too large before removal, the scar left can readily become disease infected, especially later in the season; side-shoot scars are a very susceptible entry point for botrytis spores.

Removal of Lower Leaves

This should start when the leaves have outlived their usefulness, this being indicated by their yellowing. Lower leaves can also serve as a source of fungus disease, particularly when the plant begins to hang with fruit and the lower leaves trail on the ground. It is usual merely to remove the foliage up to the ripening truss, and obviously one of the main benefits of leaf removal is that air is allowed to circulate through the base of the plants. Some growers remove leaves ruthlessly, leaving only 3ft of unleafed stem. Pruning of plant leaves has certain physiological effects on the plant, one of which is the hastening of ripening, although if pruning is carried out to excess the fruit may be smaller and quality could suffer, although there are no hard and fast rules about this.

The best way to remove leaves is, in my opinion, with a quick up-and-down movement when the plants are well supplied with moisture —certainly not in the heat of the day when their moisture content is at its lowest, and when leaf removal might result in disease-prone tears or snags. Painting any rough snags with a fungicide is an excellent preventive against disease infection.

Stopping

Opinions differ on the virtue of restricting the heights of plants to a given level. Usually this is dictated by the height of the top wire, unless the plants are to be arched or layered (see fig 24 on page 154). Stopping the plants usually takes place when there is little hope that any further fruit trusses allowed to form will ripen before the plants have to be removed, either because the season is coming to a close or in the interests of another crop.

Dutch research workers feel that it is better to allow the plants to continue growing and merely remove the flower truss, as this does not in any way check the upward flow of nutrients. It appears, however, that stopping does not restrict the upward flow, as any side shoots allowed to remain at the top of the plant develop very quickly. One reason for stopping is to avoid a clutter of leaves on the roof of the greenhouse and it could be said to be necessary on purely environmental rather than physiological grounds.

In practice all that is required is to cut or snap off the top of the plant immediately above a leaf joint, avoiding the removal of any flower or young fruit truss which has a chance of ultimately producing ripe fruit. In amateur houses with low heat levels it is unlikely that trusses in flower after the middle of August will become fertilised and produce ripe fruits on the plant, although with good heat and light levels fertilisation can occur later—well into September—producing ripe fruit as late as November.

11

Seasonal Growing Procedures

The so-called 'blueprint' growing which is the yardstick for commercial production would seem at first sight merely to consist of adhering exactly to a prescribed growing programme. Matters are, however, more complicated than this, and it is the intelligent manipulation of all the various growth factors which brings success. While the automatic aids which are becoming commonplace, both in commercial and amateur circles, take much of the guesswork out of things, there are still a great many decisions to be made, not only on the whole growing programme, but on a day-to-day basis as occasioned by changing climatic patterns and plant response to environmental variations.

GROWTH FACTORS REQUIRING ATTENTION

A great difficulty with many amateur gardeners is their absence at business for five or six days a week, leaving the plants, even with certain automatic aids, to cope with further variables. The following sections deal with the most important issues requiring constant attention.

Young tomato plants, as they become successfully established, start to show a freshening of the growing point, resulting in extension growth and the production of new leaves. Flower trusses are of course present *in* the plant before planting, although not always visible or in actual flower. The vegetative development of the plant, including flower-truss formation, subsequent setting and fertilisation, and swelling of fruit, is entirely dependent on light, temperature, water and nutrient supply, and the inter-reaction of these with each other. The concentration of CO_2 (carbon dioxide) in the atmosphere is also an important factor in development, owing to the effect this has on the velocity of the photosynthesic process.

Moisture Supplies

Water supplies should be controlled as carefully as the method of application allows. Plants supplied with an excess of water tend to go lighter in colour, and yellowing of the bottom leaves frequently occurs due to exclusion of air from the growing medium and the reduction in nitrogen supply. Plants short of water, apart from drooping or wilting badly on hot days, tend to go darker green in colour, and leaves may also go hard and brittle as plants cannot assimilate nutrients, not only because of the actual lack of water as a conveying medium, but also because of the raising of the salt content of the soil (see below). It is of some importance to consider water supplies in conjunction with nutrient supplies, as the two are virtually synonymous.

Nutrient Supplies

Nutrient element	Excess	Deficiency	Comments
Nitrogen	Lush light-green excessively curly top leaves early in season. Thin tops later in year, coupled with long internodes. Trusses may partially form or alternatively be very profuse and gross on thick stems and subsequently fail to flower or set properly. In extreme cases this will persist throughout the season, resulting in poor yields.	Poor sickly looking plants with thin hard stems; yellowing of lower leaves may also occur in severe cases.	It is important to maintain a balance between nitrogen and potash, as shortage of one can result in excess of the other and vice versa.

Nutrient element	Excess	Deficiency	Comments
Potassium	Dark-coloured leaves but stems with very short internodes and curled leaves throughout the whole plant. Flower trusses very squat. A great excess of potash will produce scorching of leaf tips (see also note on osmotic feeding, page 162).	Much the same symptoms as with excess of nitrogen, with marginal leaf scorch occurring in extreme cases.	The link between nitrogen and potash balance is a vital one with tomato plants.
Phosphorus	Difficult to diagnose visually and does not readily give rise to problems, as surplus phosphorus tends to become unavailable.	General lack of phosphorus in soil will tend to result in unproductiveness and delayed flowering and fruiting. Temporary lack occasioned by coldness causes blue colouration of leaves.	An element which is not very mobile in the soil, which explains why symptoms of excess or shortage are not so clear as with nitrogen and potash. Apart from this the tomato plant's need for phosphorus is not great.

Note: see Chapter 12 for details of other mineral deficiencies and excesses.

Light Intensities

Poor light, as occasioned by district or weather pattern, will result in thin plants of a pale green colour, with long internodes (the distance between the leaf joints). The height of the first truss will also be affected and successful fertilisation can be difficult, especially if there is industrial contamination which stains the glass and further restricts light level. Good light results in a sturdy dark-green short-jointed plant which should produce good full fertile flowers of good colour, capable of setting provided there is a correct nutrient balance. The difficulties of good fertilisation early in the season are largely the outcome of poor light, although there are certainly nutritional complications (see below): the practice of using hormone setting liquid is fairly widespread in difficult situations. It is essential that these setting hormones are used strictly according to directions and not to excess, otherwise distorted poor quality fruits may be produced.

Temperatures

Low temperatures *day and night* will generally slow up rate of growth. High daytime temperatures in good light followed by cool nights will result in a plant with curled leaves and short internodes and there can be gross truss formation and bad setting owing to the excess accumulation of carbohydrates. High night and low day temperatures will produce 'played out' thin plants with long internodes. The warmer the soil, up to about 75° F (24° C), the greater the rate of root development, although a temperature of around 65° F (18° C) is ideal. Air temperatures should be adhered to as strictly as possible (see page 148).

SEASONAL FEEDING PROGRAMME

Few tomato plants are planted into soils of low nutrient content, and it has been explained in Chapter 8 why in fact such a procedure is not desirable (see also below). While the broad approach to nutrition can be made to sound relatively simple, it must be appreciated that there are considerable complications which can arise in both soil and, more particularly, in soil-less media; with the best will in the world it is impossible to approach tomato nutrition in such a precise manner that an automatically foolproof system of feeding is possible. Conditions are bound to vary very considerably between soils, and apart from this there are temperature and light patterns, water quantities and countless other variables. It must be appreciated, however, that visual appreciation of the progress made by the plant is essential, even if the grower has ready access to laboratories for seasonal checks.

Osmotic Feeding

For the process of osmosis to be effective there should be a difference between the respective osmotic pressures in the plant and the soil (see page 51). Assuming that the plants are set out in soil or growing media which is known to be of an acceptable soluble salt level, the plants are developing their root system in the midst of a compatible chemical environment. The vital issue is the strength of the solution of salts, as this determines the rate at which the plants assimilate their moisture and nutrients, which in effect means rate of growth. Translate this salt solution into more practical terms—the available amount of nitrogen and potassium and other salts—and one moves a step nearer the whole conception of nutrition. The rate of growth of a tomato plant in a warm greenhouse is tremendous, and vast demands are made on the store of nutrients in the soil. Unless one is able to keep the soluble salt strength within the prescribed concentration and at the same time ensure that the requisite amounts of nitrogen and potash are there in correct ratio to each other, the whole system breaks down. Matters are not quite so bad as they sound, of course, as the micro-organisms in soil-based media will do much to keep the balance of nutrients stable, but this unfortunately is not the case with soil-less media.

Practical Approach to Nutrition

Assuming that there is a liberal supply of nutrients present in the first case, the next step is to estimate, both by visual means and by previous experience, whether the plant's diet is suitable. The following is the generally accepted nutritional pattern:
1. *High potash* feeds are required early in the season, especially in heat sterilised soil, and with vigorous varieties.
2. *Standard* feeds are required during the major part of the season.
3. *High nitrogen* feeds are required later in the year, when the soil re-

serves of nitrogen tend to be played out and the plants require a boost to encourage vegetative development.

It rests entirely with the grower or gardener, however, to assess the needs of his plants and to apply the necessary balance of feeds according to the symptoms exhibited by the plants, and there can obviously be no pre-set pattern in the face of all possible variables (see also page 51).

| | | | | Nutrient (% weight/vol) | |
| | **Composition of feeds (oz/gal)** | | | | |
K_2O to N ratio	Potassium nitrate	Urea or Ammonium nitrate		K_2O	N
3:1 (High potash)	24	—	—	6·3	2·0
2:1 (Standard)	24	5	6	6·3	3·3
1:1 (High nitrogen)	24	16	20	6·3	6·3

Note: this is the stock solution prepared by careful mixing in warm water; it is diluted to 1 in 200, generally with a calibrated dilutor. When using a watering-can the dilution is 1 fluid ounce of the stock solution in $1\frac{1}{4}$ gallons of water.

Proprietary Liquid Feeds

There are many proprietary feeds available and it is necessary to use these according to directions. Exactly the same basic principles apply; make sure the prescribed dilutions are adhered to and that there is the correct selection of nitrogen to potash ratio.

Solid Feeding

Most proprietary fertilisers, straight fertilisers, or self-formulated mixes can be used, adhering again to directions in the case of proprietary types, or the recommended rates for straight fertilisers. Most proprietary solid type fertilisers are used every 10–12 days at 1–2oz per square yard (usually about 1–$1\frac{1}{2}$oz).

Note that some solid fertilizers are particularly concentrated and that roots and stems may be damaged irreparably by excess or careless application. Solid fertilisers cannot be absorbed by the plant roots until they are dissolved, which means watering in or, alternatively making sure that they are applied to a damp soil. Care should be taken to avoid 'swirling' of recently applied fertilisers, which may result in concentrated patches.

Constant Versus Intermittent Feeding

What has already been said about the osmotic process will, I feel, do much to clarify precisely why it is desirable to have a constant and evenly diluted solution of fertilisers. Intermittent feeding with either liquid or solid fertilisers will obviously result in a varying salt solution in the soil, but not necessarily to a degree which will upset the plant greatly, although the first symptoms of shortage of salts will invariably be blotchy ripening (see page 197). There is virtue, therefore, in never applying plain water for border cultivation, but in the present state of knowledge it is doubtful whether constant feeding is desirable for systems involving limited amounts of growing media, especially if these are peat based. Experience has shown that a rapid build-up to a toxic level can occur where constant liquid feeding is carried out in peat-based media, and this indicates the desirability of flushing out with plain water fairly frequently.

There is a clear set of rules to follow for the accepted osmotic procedure and these are detailed on page 165. It is important to get the balance of nitrogen to potash as exact as possible.

Take care *not* to apply the nitrogen or potash feeds in excessive concentrations, whether dealing with either liquid or solid applications. With soil-less media watch particularly for any symptoms of nutritional unbalance such as scorching or bleaching of growing points, along with 'black bottoms' on the base of fruit (see page 197); these can be taken as reasonable evidence of damaging salt concentrations and, in particular, of the shutting off of calcium (see Chapter 12). Flush out with plain water when this occurs and start feeding again with the appropriate ratio of nitrogen/potash as indicated by the general development of the plant. Where improvement does not occur almost immediately, consult an advisory officer.

Summary of Feeding Programme

Liquid fertilisers: potash nitrate/urea or ammonium nitrate; 1–200 dilution. *High potash* early in season, *standard* when plants show steady growth, *high nitrogen* later in season (see page 163).
Proprietary liquid feeds: use as directed, adhering rigidly to instructions.
Solid feeds—straight: apply at rates shown in table on page 65 every 10–12 days.
Solid feeds—proprietary: apply at recommended rates every 10–12 days.

Practise constant liquid feeding early in season and continue with more restricted feeding later, provided no symptoms indicative of salt concentration trouble develop. Look particularly for blotchy ripening of fruit (indicative of low salt concentrations in the soil, and an important factor, especially in soil-less media), or 'black bottoms' (indicative of high salt concentration) and flush with water in the latter case. Where complicated symptoms occur, call in an advisory officer. Total water requirements should relate to the table on page 112.

Foliar Feeding

There is great interest these days in foliar feeding, the plants absorbing the nutrients directly into their leaves, therefore by-passing the normal root uptake. It is particularly useful for applying magnesium and calcium (see page 195) to avoid the 'shut-off' complications which generally cause these deficiencies in the first case. Osmotic principles, however, still apply, and excess concentrations of foliar feeds will cause damage to the leaves.

A quick response is generally achieved by foliar feeds. Special foliar feeds are available, but most of the normal liquid fertilisers applied by overhead spraylines will serve admirably. Some growers allow the fertiliser some 20–30 minutes to be absorbed into the leaves and then flush off with plain water to avoid any damage which could be caused by the nutrients concentrating in localised areas of the leaves in hot sun. It is worth noting that surplus foliar feed which is not absorbed by the foliage and which runs off is eventually taken up by the plant roots.

Mulching

The practice of mulching is not carried out as widely as it was many years ago, when the application of well-rotted FYM to borders in May or June was routine practice. Peat is much more widely used as a mulch today. There are several reasons for mulching:

1. To avoid surface evaporation of moisture.
2. To encourage the formation of surface adventitious roots.
3. To provide a ready source of nutrient in acceptable form to the surface feeding roots.

Mulching can be used in border culture, ring culture and, although rarely carried out, for other cultural systems. If FYM is used it should be *very well rotted*, otherwise damage can result due to ammonia release. Peat is much more palatable to the plant if approximately 5–6oz per bushel of ground limestone is added. Fertiliser can also be added to the peat, although it is better to wait until roots are developed before applying fertilisers or liquid feeds on the basis of the normal feeding programme. Peat mulches are especially useful where there is some root or vascular tissue damage due to a fungal agency or in the case of severe potato eelworm attack (see pages 178, 187).

CULTURAL MODIFICATIONS FOR OTHER SYSTEMS

Ring Culture

It has been explained that initially the roots are contained in the ring and at this stage the plants are watered and fed as described for border culture. It will be found that more water is required as the containers dry out rapidly, and I feel it would be inadvisable to apply constant liquid feeding in this instance, otherwise there could be a salt concentration problem. The roots very soon grow into the aggregate, evidence of this being the mass of fine root hairs in the vicinity of the rings, and before long the root system in the aggregate can be extensive. Early in the season all the water needed should be applied to the rings,

Page 167 The process of tomato grafting. (*Above*): The stems are cut.
(*Below*): The stems secured with transparent adhesive tape before potting

Page 168 *(Above):* A heavy crop of tomatoes, variety MG, grown in a chemically sterilised soil in the author's greenhouse. *(Below):* A rotary drum steriliser, excellent for soil pasteurisation on a fairly large scale

and the roots which form in the aggregate will be supplied by the surplus draining out of the rings. It is unnecessary and inadvisable to apply extra water to the aggregate early in the season, other than that applied for damping-down to raise the humidity for fertilisation purposes. *All* feeding is initially given through the rings.

As the season progresses *water* is given to the aggregate, while keeping the rings moist, and then water is progressively concentrated in the aggregate to encourage a gradual takeover of functions. The rings will gain all the moisture they require by capillary (or upward pulling) action. Take care, however, not to allow complete drying out of the rings until the season is well advanced. Feeding is applied to the aggregate when it is seen that the takeover is more or less complete, this generally occurring when the plants are about 5–6ft tall. Any wilting that takes place indicates that the plants are still depending on the rings. The root restriction early in the year is extremely conducive to fruit production and thereafter, when strong growth is needed, the roots in the aggregate are capable of sustaining growth throughout the season. Support, training and general cultural procedure is similar to that for border culture.

Grafted Plants

A slow start is typical of grafted plants, especially if root disturbance has occurred when the fruiting variety root is removed at planting. Once plants do get under way, however, they develop great vigour and cultural procedure is similar to that described for border culture.

Straw Bales

The need to maintain water supplies is the main problem with straw bales, and some 'automatic' watering system such as drip nozzles or perforated hose is desirable, coupled with draping the bales with polythene, leaving only a section of the soil ridge exposed. There can be 'locking up' of nitrogen early in the season due to the concentration of bacteria on the straw decomposition, thus necessitating the application of extra nitrogen; thereafter there can be a surplus of nitrogen, necessitating liberal potash additions, especially to keep fruit quality high.

Other Systems

Most of the problems are nutritional and have already been referred to in some detail. The physical application of water creates problems for some systems, unless drip or other methods of watering are adopted.

Important Note

The frequency of applying liquid fertiliser with all systems involving limited quantities of growing media must be the subject of constant check, especially where there is excessive moisture loss, as with ring culture, straw bales, or polythene bag culture, if salt problems are to be avoided. Reference back to the total quantities of nutrients required by plants as detailed in Chapter 5 is advisable, but constant visual assessment is still of prime importance.

FRUIT PICKING TO END OF SEASON

Picking tomato fruit is a relatively simple task, the main provision being that the fruit is removed by snapping the stem and preferably not by removing the fruit from the calyx. It is important to handle fruits carefully and keep them cool, especially in hot weather. Leaving the fruit lying about in the greenhouse, or even in the sun out of doors, is certain to induce softening. The stage at which to pick the fruit depends on whether it is to be consumed immediately or whether it is to be sold; it should be noted that where *wholesale marketing* is involved, compulsory grading standards apply, and details of these are available from any advisory or marketing officer (see Appendix 2). Fruit intended for wholesale markets is usually picked fairly green.

Crop Finishing

Reference has been made earlier to the nutritional pattern necessary for sustaining growth, usually by varying the nitrogen to potash ratio, ranging from high potash early in the season to high nitrogen late in the season. It is important to keep a strict watch on the general vigour of the plants and apply the corrective treatment at the right time as the real success of the tomato crop depends on the overall performance, not merely on the early crop.

'Fade-out' of plants due to a variety of causes can frequently occur and these are detailed in Chapter 12. Maintaining a healthy vigorous root system right to the end of the season is not always easy in older borders which have been in cultivation for many years. Despite sterilisation, root rots and vascular diseases can quickly reduce end-of-season vigour, as indeed can virus infection or eelworm infestation contracted from deeply seated sources. This is in fact the main reason why systems other than border culture are finding such favour.

End of Season Watering Patterns

These can be confusing as, while a plant in full vigour will use the calculated or estimated amount of water (see page 112), a heavily deleafed plant, especially if suffering from root or stem disorders, is unlikely to assimilate large quantities of water. It must also be remembered that even in the case of a plant with a healthy root system some natural root death occurs. Nevertheless, as the season progresses the main roots can become fairly extensive and the plant is able to draw on the reserves of water in the soil at lower levels. This explains why, despite the cessation of regular watering which often accompanies the general lack of gardening enthusiasm at the end of the season, the plant continues to grow strongly. There is in fact some virtue in restricting water quantities at the tail-end of the season, as it definitely encourages the ripening of remaining fruit.

A further point of importance is that the high humidity of the atmosphere which often occurs naturally out of doors in August/September/October gives rise to a humidity problem in the greenhouse, encouraging disease. The excess application of water raises the humidity pattern still higher, and botrytis can become a major problem. The application of less water at this late stage in the season will help to keep humidity down and coupled with this precaution, heat should always be given at night.

Removal of Plants

When it is decided to remove the plants, all the remaining fruit is picked and either ripened at a light window or on a greenhouse bench. Green fruit can of course also be used for making chutney. The plants are cut 9–12in above ground level and detached from their strings, after which they are dumped some distance from the greenhouse. Alternatively they can be destroyed by burning as soon as dry enough. The plant stumps are then gently forked out of the soil with as much root as possible, watering the border to facilitate this if necessary. The plant roots should also be disposed of away from the greenhouse, or burned. Time should be taken to examine the roots carefully to estimate degree of fungal or pest attack.

Greenhouse Cleansing

A completely empty greenhouse can be cleaned out immediately after crop removal, although it is usual to wait until nearer the time of use again, as the glass can quickly become very dirty. Fumigation can, however, be carried out immediately by burning sulphur at 1lb per 1,000cu ft, though this is a dangerous practice for amateurs unless strict precautions are observed. Formaldehyde or cresylic acid can be used for washing down after fumigation, any time up to five or six weeks before the house is to be used again. It is put on with a sprayer, taking care to protect the eyes, and then hosed down after a week or so. Alternatively a good detergent or mild sterilant can be brushed on with a long handled brush and hosed off subsequently, trying to forcibly remove as much moss and algae as possible.

Note on Yields

Tomato yields differ enormously according to climatic pattern, variety, environmental conditions, planting density, situation, incidence of pests and disease and countless other factors. Commercial yields vary from about 40 tons to over 90 tons per acre, which is approximately 6–15lb per plant. A good average yield would be between 8 and 10lb per plant.

12

Diseases, Pests, Physiological and Nutritional Disorders

The tomato plant is unfortunately prone to a wide variety of troubles, many of which appear to assert themselves with some regularity. Undoubtedly the principle of growing the same crop successively in the same greenhouse, and often in the same soil, is partly responsible for this. This is why systems of culture involving fresh growing media are becoming so widely practised, although even with a clean start there are other troubles which are still persistent, no doubt because of the 'artificial' conditions in any greenhouse where plants are grown in limited quantities of growing medium. A plant with so many vital facets of growth requires a considerable degree of precision in culture, and obviously it is all too easy to relax attention.

Having painted this rather gloomy picture, I must also point out with some haste that certain troubles are so well known that it is relatively simple to take evasive or preventive action. For example, it is known that planting in cold soil checks the roots, allowing easy entry of parasitic fungi. Likewise it is readily appreciated that a highly humid atmosphere will induce the attack of botrytis (grey mould). Sufficient is also known about nutritional programmes for the grower to realise that a departure from the correct treatment will have repercussions. Plant breeders have done much to increase resistance to diseases, and this applies especially to leaf mould, vascular wilts, and virus disease (TMV).

A catalogue of troubles must now necessarily follow, but at all times preventive measures will be emphasised.

One final point which must be made is that it is folly to try and take short cuts, or to imagine that if, for example, a root rot occurred severely in a border soil one year, it will vanish merely with the passage of time, some lime, manure and fertiliser. Organic principles of husbandry apart, which is a massive subject involving detailed study, I feel that it is as well to face up squarely to problems and sterilise effectively (see page 200), re-soil, or go on to an alternative cultural method.

Very Important Note

In the course of dealing with the control of pests and diseases, reference will be made to various chemical preparations. It cannot be stressed too strongly that many of these chemicals are extremely potent poisons and *must* be used strictly as recommended, and kept safe in a locked cupboard. The active chemicals involved are stated in this book, and the trade names can be found by referring to up-to-date copies of the HMSO booklet *Agricultural Chemicals Approved Scheme—Approved Products for Farmers and Growers.* While this is intended mainly for use by professionals, many of the chemicals listed in it are available to amateurs. Commercial firms also issue lists of their products for both professionals and home gardeners. Some confusion exists regarding the supply of many chemicals and while professional growers will have little difficulty in obtaining any referred to, amateur gardeners are advised only to use materials freely sold in retail establishments.

While reference will be made to certain 'established' chemicals in the course of the book, readers should refer to Appendix 1 for the current up-to-date chemicals (1975) listed with their appropriate role, extracted from the booklet previously referred to.

Diagnostic Techniques

Plant pathology, entomology, and physiology (including nutrition) is a highly specialised subject requiring considerable detailed knowledge and, in many cases, highly complicated equipment. There are therefore obvious limits to what can be correctly diagnosed merely by the recognition of relatively standard symptoms. Nor is it the case that the symptoms exhibited always follow a single pattern or, for that matter, that only one trouble may affect one plant at a time. There would appear to be little rhyme or reason why some plants are attacked and others remain unaffected, although they may be growing in close proximity.

On the other hand, trouble may be relatively simple and clear cut, and the use of a high-power magnifying-glass may reveal much. Spore forms may also be remarkably distinctive and reference to a book on pathology will give good guidance. Where, however, there is any doubt about a malady, it is wrong to act impulsively and attempt a cure which could do more harm than good, and here it would be very wise to consult your local horticultural adviser, who is invariably backed by a team of specialists.

DAMPING-OFF

Symptoms. Young plants fall over when 1½–2in high, the stem at soil level being shrivelled or 'pinched' in. Seedlings may seem unaffected only to collapse at a later stage with root rots as a secondary infection.
Agent involved. Fungal parasites, phytophthora, pythium, corticium and others gain entry either through the roots or the stem.
Prevention. Avoid sources of infection such as dirty seed boxes, pots, soil, etc. Clean fresh peat, peat/sand mixes or properly prepared (sterilised) John Innes Seed compost should be used, although the ammonia content of a soil-containing media can injure the plants and afford easy entry for fungi, indicating that a soil-less medium is desirable. Here again, however, ammonia release from fertilisers such as hoof and horn can still be a problem. Dirty water can also carry infection and static tanks are a potent source of trouble.
Control. Copper-based compounds or Cheshunt Compound are useful in preventing or limiting the spread of attack, but will not greatly help plants already attacked. Excessively high humidity should be avoided after the actual germination period is over. Where greenhouses are lined with polythene there can be damping-off problems unless ventilation is effective. The fungicides should be used according to directions.

ROOT DISORDERS

These attack portions of the plant below and at ground level, ie the root is affected and the uptake of water and nutrient is restricted according to the inroads made by the parasitic fungus involved. This could be one of a number of different species including colletotrichum, atramentarium, thielaviopsis basicola, rhizoctonia, pythium, phytophthora, didymella, or fusarium. While accurate identification of the fungal parasite involved can be made by a pathologist, it is obviously difficult to be completely accurate by visual identification only. Any form of root rot will cripple the growth of the plant to a greater or lesser degree, and identification of the parasite is of largely educational interest—including, of course, prevention and control for other unaffected plants or for the future.

While a certain amount of control may be exercised on root rots, it is difficult to gain access to the fungal infection in the root tissues without damaging the plant cells. Systemic fungicides are now being used for certain aspects of disease control and these may offer great scope for the future.

Types of Root Rot

Root rot. Fine rootlets at or near soil level are attacked by the disease organism, leaving the main roots devoid of side roots or root hairs. The outer 'skin' of the root is usually attacked, allowing it to be removed, leaving behind the wiry vascular tissue. Intensity of attack varies considerably and only one section of the root may be attacked. A plant affected by root rot may continue to grow reasonably well, although wilting will occur when the temperature rises or when it changes quickly. The portion of root so far unaffected may in fact also continue to develop and to compensate in part for the damaged portion, and this state of 'brinkmanship' may exist for the greater part of the season. Alternatively the progress of the disease may be rapid, as indicated by complete collapse of the vegetative portion of the plant. Damage of any kind to the root tissue—cold soil, over- or under-watering, excess fertiliser, or the nibbling of pests such as wireworms, slaters or eelworm —obviously destroys the outside protective tissue of the root, allowing ready entry of the disease organisms.

Foot or collar rot. The disease attacks at soil level and quite frequently the attack is connected with physical damage of some sort, including burning by excess fertiliser application. The stem at ground level will show a brown ring of tissue with shrivelled stem, and collapse of the plant could occur when sufficient inroads have been made into the vascular tissue.

Toe rot. The extremities of the root are affected and, as with root rot, the restriction in growth is related to the severity of the attack. This is a frequent trouble arising at planting time, due to cold soil at lower depths. New adventitious roots will frequently be seen, initially forming as little lumps on the stem above soil level. Once these roots extend into the soil they may carry the plant on for a time, although collapse later in the season usually occurs.

Brown or corky root (sterile grey fungus). This is caused by the fungus *Pyrenochaeta terrestris* resulting in a clubbed and brown appearance of the root, reducing its efficiency very considerably. It is frequently present in old tomato borders and very difficult to eliminate—even by sterilisation.

Prevention of Root Rot

Use of 'new' or properly sterilised soil is essential, especially if a previous crop has been seriously affected by one or other of the root rots, as spores or infected debris will remain in the soil and affect succeeding crops. 'New' soil means soil which has not been used for tomato (and preferably potato) cropping for a number of years, and there is every likelihood that such soils, provided they come up to the necessary physical and nutritional standard, will give excellent results, free of root rot, but there is no guarantee that this will be the case. The reason is that root rots (and for that matter other diseases) while relatively specific to tomatoes, are not entirely so, apart from which infection can frequently be air- or water-borne.

Sterilisation methods are dealt with in Chapter 13, yet unfortunately it seems very difficult to get rid of certain root rots (and other troubles), especially brown root, unless sterilisation is 100 per cent efficient, and this is hard to achieve in practice. Nevertheless, sterilisation techniques can on the whole be considered as effective, and where carried out correctly there is every likelihood that cropping results will be good, *within the limits of the sterilisation process employed.*

Control

Plants affected by root rots can be encouraged to stay alive if the progress of the root rot is checked, or limited to the affected area; and at least by using control methods the incidence of disease may be limited to as few plants as possible. Contact fungicidal chemicals such as dithane (nabam) and Sterizal are frequently used and it is likely that others may be available shortly following the success of the systemic fungicide Benomyl for botrytis and other diseases (see below). In all cases use the chemicals as instructed. Preventive watering with these chemicals where root rots are suspected may also be helpful.

Spreading a layer of peat and lime (5–6oz ground limestone per bushel), 'clean' soil, leaf mould, or other compatible material will invariably induce the tomato plant to produce new adventitious roots at or above soil level, and these roots will sustain a plant suffering from some root debility for a period at any rate, depending on the severity of the attack. A plant very badly affected by root rot may recover only very briefly, if indeed it manages to produce adventitious roots at all, before complete collapse of the root system occurs. On the other hand, mulching may help to sustain the plant sufficiently to produce a worthwhile crop, a lot depending on whether the new roots are able to cope with high temperatures if there is any very hot weather.

Combining mulching with shading is a frequent practice, painting the *outside* of the glass with lime and water or with a proprietary shading material. I have found highly diluted emulsion paint useful. If the ventilation is reduced for a week or so and frequent damping-down practised, the excessively humid conditions which develop will reduce the transpiration of the plant and encourage adventitious rooting. Adventitious roots once formed will of course respond to nutrient application just as 'real' roots do, provided the mulch is kept moist.

Mulching does in fact reduce moisture loss from the soil surface and can be carried out irrespective of whether or not there is incidence of root rots. In this case, however, straw is frequently used, which, in addition to reducing moisture loss, keeps the soil from splashing on to the plants.

VERTICILLIUM AND FUSARIUM WILTS

The fungal agent enters the plant usually through the roots and affects the vascular tissues, clogging these up by its exudations. Affected plants, usually 4–5ft tall and laden with fruit, will wilt when the temperature is sufficiently high to cause rapid transpiration, and frequently recover under cooler conditions. Cutting off the stem at ground level, then making a longitudinal cut along the width of the stem tissue, will show quite clearly the brown area.

Verticillium wilt is a disease more associated with low temperature conditions in the north, while fusarium is more common in, but by no means confined to, the south. Apart from wilting, the leaves drop badly, accompanied usually by yellowing and withering of the lower leaves, often only at one side of the plant. Soils frequently build up an infection of verticillium or fusarium, which asserts itself in patches and spasmodically, and usually when the season is well advanced. Root rots can frequently occur in combination with wilts.

Prevention

The use of 'clean' or sterilised soil is essential, otherwise alternative cultural methods should be adopted. Avoid low temperatures and planting checks. Use grafted plants on root stock KVNF, the V and F referring to verticillium and fusarium respectively. Varieties are also being produced with inbuilt resistances (see Variety List pages 210–13).

Control

Once plants are seriously affected not a great deal can be done to alleviate the situation. Very badly affected plants are better removed. Indeed root rots frequently occur in badly debilitated plants in any case. Mulching, as described above, accompanied by about fourteen days at 77° F may enable plants not only to survive, but in the case of verticillium wilt to 'cook out' the disease organisms involved. Benomyl has given encouraging results in the control of wilts.

DISEASES AFFECTING THE STEMS, LEAVES AND FRUIT

Two major diseases attack the stem of the tomato—didymella and botrytis—and both can cause immense damage.

Didymella (*Didymella lycopersici*)

This attacks the stem of a mature plant a few inches above soil level, causing brownish lesions dotted with shiny spores, which can be seen if examined with a lens. Leaf spotting can also occur, and in addition the fruit stems may rot, causing the fruit to drop. Damage ensues when the mycellium of the fungus cripples the vascular tissues. In a heavy attack there can be a tremendous toll of plants, yet there are occasions when only a few plants are affected. Infection arises from spores, and from contaminated seed boxes, soil, string, structure of the greenhouse, base walls, or anything which has been in contact with infected plants. Didymella generally tends to be spasmodic, although in certain areas it is a constantly recurring problem.

Prevention. Immediate *complete* removal and *burning* of infected plants, coupled with thorough spraying of the remaining plants with captan or dithane, concentrating the spray on the lower stem areas (other more potent chemicals were used for a time, but have now been withdrawn). Thorough hygiene of a routine nature is essential, particularly following an outbreak, and this includes burning sulphur and thoroughly washing down the greenhouse structure, coupled with thorough soil sterilisation or the adoption of alternative cultural methods. It should be noted that even if this involves 'new' growing medium, infection can still arise from other dirty sources. It is important that any doubtful plants at the end of the season be disposed of at once at some considerable distance from greenhouses.

Botrytis Stem Rot (*Botrytis cinerea*)

The spores of grey botrytis mould abound in the atmosphere and gather readily on any receptive host. In the warm humid atmosphere of the greenhouse botrytis spores are in their element and it merely requires a dropped petal, the open scar of a removed leaf or side shoot, or rotting lower leaves, for the spores to germinate and start plundering, at a rate highly dependent on the humidity of the atmosphere and the softness of the tissue.

When air movement is restricted in any way, either by poor ventilation or dense foliage, conditions are ideal for the development of the fungus, sending its mycellium into the tissue and sporing profusely, as indicated by the mass of grey dust-like spores which develop.

Botrytis also causes damage to the fruit by 'ghost spotting'. This happens when, in conditions of inadequate ventilation or low temperature at night, a droplet of moisture laden with spores falls on the fruit, and starts to attack the skin; it dries up again during the day—but obviously the fruit can become so badly marked that it loses its fresh appeal, and of course its sale value. The fruit is also attacked at the husk, causing it to drop, and flowers also are frequently attacked, rendering them useless.

Prevention. Adequate ventilation and the avoidance of high humidity, especially at night (by ventilation *and* the use of night heat), timely defoliation, good balanced nutrition to keep plants hard rather than soft (some varieties have a considerably softer foliage than others, eg Moneymaker), removal of all decayed leaves and debris. Fan ventilation has proved excellent for the prevention of botrytis because it can give positive air changes regardless of all weather conditions, but a time clock is required to give periodic night operation. Good hygiene pays big dividends: fallen leaves and other debris should never be left lying around.

Control. Mild attacks by botrytis (before the disease makes massive inroads) can sometimes be effectively dealt with by scraping the stem down to clean tissue, and provided environmental conditions in the greenhouse are good and that some material such as lime and flowers of sulphur is used to dry the tissue, such control can be effective. Captan and copper compounds are useful, failing which dicloran dust and TCNB can be used to very good effect, TCNB being available as an aerosol in smoke form. Even creosote can be used, this being painted on in neat form to the actual infected area. Great success with benlate (Benomyl), which is a systemic fungicide, offers great hopes for the future. Dichofluanid (Elvaron) has given good results against ghost spotting and botrytis.

Leaf Mould Disease *(Cladosporium fulvum)*

This at one time was an accepted part of tomato culture: at various times of the season, usually after about June, yellow patterns appear on the upper surface of the older leaves, while on the underside felt-like patches of brown fungus develop. The efficiency of the leaf is obviously impaired and the disease usually progresses until the whole plant is enveloped apart from the top portion which generally remains relatively clean. (This disease has a special personal significance for me as I became asthmatic when I entered a greenhouse where cladosporium existed in any quantity, an unfortunate allergy of which I was eventually cured by a course of injections). Fortunately many varieties of tomato are now resistant to A and B strains of leaf mould. In recent years a further strain (C) of cladosporium has developed, and this appears to be becoming a problem as some varieties have no resistance to it. The situation regarding leaf mould resistance is constantly changing, due to mutations of the disease. Mild damp areas, especially those on the western seaboard, favour the development of this mould.

Prevention. Obviously the use of resistant varieties is the first line of defence, remembering the existence of different strains of cladosporium (A and B, and now C). Adequate ventilation day and night, with heat at night to help air movement, will frequently keep cladosporium to the very minimum especially where non-resistant varieties are grown on account of their high quality and excellent cropping potential (eg Ailsa Craig or Moneymaker—see Variety List, pages 209–17). Alternatively chemical sprays based on maneb, nabam and zineb can be used as a preventive measure, strictly according to directions. Aerosol packs of these chemicals may also be used.

Control. Mild attacks may be controlled or restricted by use of the chemicals referred to. Removal of the worst affected leaves before application is also helpful.

Sclerotinia Stem Rot (*Sclerotinia sclerotiorum*)

The base of the stem is attacked and a white mould follows the appearance of black or brown lesions. The fungus causes the plant to wilt or die when it makes sufficient inroads. Careless fertiliser application or burning of the stem often initiates the attack. Black shiny spores can clearly be seen through a lens.

Prevention and control. Use a contact fungicide such as Sterizal or captan at an early stage of attack. Sterilise or change soil for future use.

Buck Eye Rot

This is typified by the formation of grey or reddish brown patches with inner circular rings, almost like the eye of an owl; infection occurs usually from soil splashing during careless watering.

Prevention and control. All affected fruit must be removed, and care and attention given to watering. Spraying with dithane will restrict the attack.

Potato Blight (*Phytophthora infestans*)

Not really a problem under glass, except in areas where a lot of outdoor tomatoes are grown. Autumn tomato crops may be infected if grown out of doors for a period before being brought under glass. Symptoms are blackish-purple marginal areas on leaves, which eventually exhibit a white downy growth followed by serious defoliation. Fruit may also be attacked, causing dark brown blotches which penetrate into the flesh rendering it inedible.

Prevention and control. The BBC's blight forecast in farming programmes from about mid-June will advise when conditions are humid enough out of doors for the blight disease to develop. Spraying with Bordeaux mixture, maneb or zineb is the stated control, and I feel that it is only necessary to do this with the outdoor crop or when plants are temporarily our of doors.

VIRUS DISEASE

Virus disease on tomatoes presents a very complicated picture, and only a pathologist using advanced techniques is able to positively identify virus diseases, although the more common virus disorders exhibit fairly clear symptoms. Virus diseases exist in the cells of the plant as ultra-microscopic rod-shaped organisms which plunder the plants' protein content and upset its physiology in some way—usually by reducing photosynthesis or causing acute distortion, stunting or blotching. Virus is spread readily by hands, knives, physical contact of plants, and by actual contact of the plant roots with affected debris in the soil. It is in fact very difficult to get rid of deep-seated virus infection in the border soil, as steaming is only usually effective to a limited depth in tomato borders. This is one reason why alternative cultural methods are practised.

Types of Virus Disease

Tomato mosaic virus (TMV). Pale yellow areas on the leaves of the plant may be accompanied by some destruction or roughening of the leaf according to the severity of the attack. Setting of fruit (the fertilisation process) may also be prevented, especially at the fourth to sixth truss stage) and there may be blotchy ripening of existing fruit. There can also be an area of dead cells around the vascular area under the skin of the fruit, resulting in 'bronzing'. The leaves of the plant may also become thin and develop a 'fern-leaf' appearance. The top of the plant may also become very hard and thin and there can be a general check to growth, again usually at the fourth to sixth truss stage.

Severe virus streak. This stunts the plants badly and spreads rapidly. Dark brown streaks appear on the stems, which become hollow. Fruit may also become blotched.

Double streak. Not normally a problem with glasshouse plants; it is more usual out of doors, when plants are badly stunted and die back.

Spotted wilt. Here the upper leaves of the plant tend to curl downwards and inwards, exhibiting a bronze colouration, and fruit may exhibit concentric brown circles. This is a common malady of plants such as dahlias or chrysanthemums and is usually transmitted initially from them by sucking pests.

Cucumber mosaic. Typical fern-leaf symptoms develop. Apart from cucumbers, this is a malady of outdoor plants, transmission being by insects.

Prevention of Virus

This is a difficult matter as the many sources of virus disease are very hard to eradicate. Obviously hygiene is of prime importance, washing down greenhouses carefully and carrying out heat sterilisation methodically. Chemical sterilisation of the soil is of little avail, as it exercises no control on virus infection contained in root debris. A typical situation is that heat sterilisation clears the top 10–12in of soil of virus infection, yet when the plant roots penetrate beyond this, and they invariably do as the season progresses, they pick up infection. Alternative cultural methods using limited quantities of clean growing medium can ensure freedom from virus in the absence of contamination by either mechanical or insect agencies.

Virus-tested seed, which has also been heat and chemically treated to free it from virus infection, is readily available. Care at the germination period will also minimise infection (see page 88) and dipping fingers in 2 per cent tri-sodium phosphate during the pricking-off operation is a useful practice. Avoidance, as far as possible, of sudden temperature changes would appear (from experimental work in Holland) to help, as the 'check' period is found to coincide with virus onset; the principle behind this is that the virus is always present and makes progress when the plant is under strain. Sophisticated control equipment has been developed to help overcome 'check' periods.

Spraying with sugar or milk was for a time practised, this being to offset the plundering of the protein by the virus organisms, but it seems that normal foliar feeding can accomplish the same results. Experiments were carried out by inoculating young plants from a cold storage virus vaccine thus bringing the 'check' period forward. Initially carried out on the Isle of Wight, there has now been considerable development of the technique. The introduction of virus-resistant and 'tolerant' varieties of tomatoes with resistant genes has raised considerable controversy owing to the rapid mutation of virus strains and the obvious long-term difficulties which this implies. The problem of growing *virus-tolerant* varieties along with non-resistant varieties cannot be over-emphasised, as the former may contain virus but not exhibit virus symptoms.

Control of Virus

Only in severe cases should affected plants be removed and burned. It is doubtful whether there is any real value in removing TMV-affected plants. The practice of applying quick-acting straight sources of nitrogen such as nitro-chalk or Nitram at about 1oz per square yard, or in liquid form at 1oz per gallon for containerised systems, is useful. The rules given for the prevention of virus will serve as control measures for the future.

TOMATO PESTS

Potato Eelworm (*Heterodera rostochiensis*)

The major scourge of potatoes, this is also a serious pest of tomatoes (which are also solanaceous). Infestation occurs either from former potato land on which the greenhouse has been built, or from infected soil brought into it. The persistence of the eelworm is due to the protection afforded to the larvae by the tough-skinned swollen abdomens of the females; these cysts, as they are called, are very resistant to outside influences. The larvae are induced to become active by stimulation of a solanaceous acid exudation (and possibly by other agencies).

Symptoms of infestation are the wilting of the plant in hot sun and general debility despite liberal feeding. Microscopic eelworm nibble at the *root tissue* (not the stem) reducing its efficiency and opening the way to fungal attack. Comparatively small eelworm attacks initially can quickly build up in severity and result in yellow patches, followed by dry dead patches, on the lower leaves. Confirmation of attack is the finding of minute cysts—white, yellow or orange according to their age—on the fine root hairs. Accurate eelworm counts can be carried out by the Advisory Services. Such eelworm counts are of course better carried out prior to using the soil for tomatoes.

Prevention and Control

Completely effective soil sterilisation of greenhouse borders is difficult to achieve owing to the presence of eelworm in the lower depths of soil or greenhouse foundations. Sterilisation by heat will, however, keep the eelworm attack down to moderate proportions. Chemical sterilisation varies in its efficiency (see page 204), a lot depending on the evenness of application and condition of the soil, but again it is usually possible to keep infestation to reasonable proportions.

Tomato grafting appears to give reasonable results, no doubt because of the vigour of the root system (see next section). The development of plants resistant to potato eelworm is a reasonable possibility for the future. Alternative cultural methods do, of course, overcome the eelworm problem, provided there are no possible sources of infestation.

Watering with Sterizal can be remarkably effective in containing the attack to reasonable proportions, and foliar feeding can also help to overcome debility by short-circuiting the roots. Mulching with peat (see page 166) can also help by providing the plant with sufficient new roots, as yet unaffected by eelworm, to sustain it. Combined D-D (see page 204) and solubilised cresylic acid treatment appears highly effective, the former chemical dealing with the lower depths of soil, and the latter with the upper layers.

Root Knot Eelworm

Several different species of root knot eelworm, all belonging to the meloidogyne family, attack and enter the roots of the tomato, causing swellings, nodules and malformation. Roots are restricted in their efficiency according to the severity of the attack and mild attacks cause plant wilting in warm weather. Major onslaughts are disastrous, with severe wilting, yellowing leaves, and death of plants.

Although not cyst forming, these eelworm exist in plant debris and are persistent. Preventive and control measures are as for potato eel-worm, watering with the poisonous chemical Parathion being prac-tised commercially only.

The letter 'N' in root stock types refers to resistance to root knot eelworm, and in practice this resistance is highly effective.

Symphylids and Springtails

These are active insects of various colours, ¼in in length, which can in-fest the roots of tomatoes, especially under warm, dry conditions early in the season. They are very mobile pests and move up and down according to soil temperature and conditions. They suck the sap of the roots causing loss of efficiency and cause distortion with their exuda-tions. They are more serious in the warmer southern counties than in the cooler north. Presence can be confirmed by taking some soil from the vicinity of the plant roots and placing this in water, when the in-sects will float to the surface.

Prevention and control. Soil sterilisation is usually effective, although re-curring infestations are likely because of the ability of these pests to move down below the level of sterilisation. Insecticidal soil drenches are reasonably effective if applied heavily (see page 226).

Red Spider Mite (*Tetranychus urticae* or *cinnebarinus*)

A crippling pest on certain soft-leaved varieties such as Moneymaker, which seems especially prone to attack. The tiny yellow or red mites concentrate on the underside of the leaves, causing acute loss of sap, speckling of leaves, and eventually a dry dirty appearance with a mass of webs. Countless generations develop in the year and over-winter in the greenhouse structure or any alternative host plants.

Prevention and control. Chemical cleaning of the glasshouse structure, particularly the base walls, after first removing and burning affected plants. Sulphur burning is advised.

A varying chemical programme is usually practised to avoid the build-up in resistance which can result from constant use of one chemical. As there are usually overlapping generations, control measures should be carried out on a concentrated basis over a short period. Sprays or aerosols (including smokes) may be used. Biological control is being experimented with.

White Fly (*Trialeurodes vaporarium*)

One of the worst and most persistent pests. Eggs are laid by the waxy white adult in circular clusters on the lower leaves, and after a nymph stage the adult emerges and feeds on the leaves, sucking sap and exuding a honey dew on which a fungus grows. Serious attacks damage plants severely and greatly restrict cropping. White fly is very difficult to control owing to the resistance of the nymph stage to chemicals.

Prevention and control. As white fly over-winter on perennial plants and debris, very thorough cleaning is necessary, including the use of sulphur. Decorative plants such as arum lilies or fuchsias over-wintered in the greenhouse are frequently hosts to white fly.

Whatever insecticide is selected, it is advisable not only to vary the treatment, but to ensure that two or three applications are given in quick succession. It is advisable to spray the soil in the morning to kill off any surviving flies. Biological control is worth considering, although there are still some problems in this direction.

Thrips and Aphids

Many different species can attack tomatoes, both at the young stage when they cause severe distortion and growth restriction, and later in the season. Aphids cause distortion with yellow patches on leaves and white blotches on fruit. They also exude honey dew. Thrips cause silvery blotches.

Prevention and control. Use of insecticides, varying the type to avoid building up resistances, especially as many types of aphids seem already to have developed resistance to many insecticides.

Tomato Leafminer (*Liriomyza solanii*)

This is a common pest, but like all leafminers, it tunnels into leaf structure leaving a trail of destruction.

Prevention and control. Remove infested leaves and burn; make frequent use of insecticides.

Tomato Moth

This can be a damaging pest, the green caterpillars nibbling at the leaves and stems, and eating holes in the fruit. Eggs are laid in batches on the underside of leaves.

Prevention and control. Good hygiene and early use of insecticides.

Miscellaneous Pests

Woodlice. The familiar flat 'slaters' attack young or older plants at the base of the stems. Special insecticidal dusts can be used, failing which place sections of turnip or potato around the plants and burn congregating woodlice, preferably with a flamegun or blowlamp.

Mice. Always a problem with any crop, they can eat seed or nibble at young plants. Trap or bait.

Wireworms. Common on new pasture land, the familiar wireworms enter the main stem and tunnel up the centre. Use insecticidal dusts or sterilise the soil.

Millipedes. Curled up like a coiled spring, these can be a nuisance if present in sufficient numbers, but can be controlled with an insecticidal drench.

TOMATO PESTS AND DISEASES CHART

Readers should consult also the chart on mineral and physiological disorders on pages 195–9 as there may frequently be a complex of conditions. A list of chemicals, compiled from the Ministry booklet *Approved Products for Farmers and Growers* appears in Appendix 1.

Trouble	Symptoms	Cause	Control
Damping-off	Young seedlings heel over at soil level.	Use of incorrectly sterilised or dirty soil or containers. Ammonia release may induce damage.	Use clean containers and soil. Apply copper-based Cheshunt Compound.
Root rots Root, foot, toe, and corky root rot	Plants wilt and may eventually die, depending on severity of attack. Roots seen to be brown and dried looking.	Infected soil or cold checks. Young plants may have had infection from propagating stage.	Remove infected plants. Sterilise soil for future. Use of fungicide in liquid form may limit attack. Mulching is useful for producing new adventitious roots at soil level.
Wilts Verticillium and fusarium	Plants wilt badly in hot sun but can recover at night. Advanced symptoms —leaves yellow and hanging. Vascular tissue seen to be brown if affected plant cut at ground level.	Infected soil.	Mulching and sterilising is only real way of overcoming attack. 3 weeks at 77° F for *verticillium only*. Use grafted plants. Resistant varieties also available.
Diseases affecting stem, leaves and fruit Didymella	Plants show diseased lesions a few inches above ground level and will wilt and die.	Infected soil.	Remove affected plants and spray remainder.
Botrytis	Stem rot. Ghost spotting of fruit. Leaves rot and fruit	Airborne infection.	Give adequate ventilation, especially at night to avoid high

Trouble	Symptoms	Cause	Control
	drops due to attack at husk.		humidity. Scrape stem clean and paint with fungicide or creosote. Use aerosols.
Leaf mould	Leaves yellow spotted with brown felt-like fungal growth on the underside.	Air-borne infection.	Give adequate ventilation, especially at night. Use fungicidal sprays. Use resistant varieties.
Sclerotinia stem rot	Base of stem attacked.	Soil-borne infection.	Use fungicidal wash. Sterilise soil for future.
Buck eye rot	Diseased 'eyes' in fruit.	Soil-borne infection.	Remove all diseased fruit and take care with watering. Spray with chemicals.
Potato blight	Grey blotches on leaves. Sometimes black areas on fruit. Crops rarely affected under glass.	Air-borne infection.	Preventive sprays with Bordeaux Mixture or other fungicides.
Virus diseases			
TMV (tomato mosaic)	Mottling of leaves. Bad setting. Bronzing of fruit.	Transmitted from soil, debris and seed etc.	Give nitro-chalk at 1oz per sq yd. Grow resistant (or tolerant) varieties.
Severe streak and double streak	Markings on stem and die-back of plant.	Transmitted from soil, debris and seed etc.	Remove affected plants and burn.
Spotted wilt	Top of plants turn down. Fruit has concentric brown circles.	Transmitted from plant remains etc.	Remove badly infected plants and burn.
Cucumber mosaic	Fern-leaf symptoms of leaves (do not confuse with hormone damage, for which see page 199).	Transmitted by sucking pests.	Remove severely affected plants and keep weeds down in greenhouse.

Trouble	Symptoms	Cause	Control
Pests			
Potato root eelworm	Wilting, blotching and yellowing of lower leaves.	Infection arising from cysts in soil that was pre-previously used for potato culture.	Sterilisation of soil. Watering with chemicals. Mulching and foliar feeding.
Root knot eelworm	Wilting and general debility. Roots affected with swellings.	Soil-borne infestation.	Sterilisation or use of root stock. Mulching. Soil drenches.
Symphylids and spring-tails	Very small insects in clusters on roots.	Deep-seated infestation.	Sterilisation. Soil drenches.
Spider mite	Tiny little brown or red spiders on under-side of leaves, causing speckling.	Arising from over-winter-ing pests in greenhouse.	Chemical sprays or aerosols, applied over concentrated period. Biological control.
White fly	White flies feeding on leaves, exuding honey dew which develops brown/black fungal growth.	Over-winters in vegetative tissue or in cracks.	Sprays or aerosols, over concentrated period. Biological control.
Thrips and aphids	Various types of greenfly and sucking pests, all sucking sap and causing general debility.	Infestation arises from various sources, fre-quently from outside crops.	Sprays and aerosols, applied over concen-trated period, varying the chemicals.
Tomato leafminer	Tunnelling of leaves	Infestation usually im-ported from outside crops.	Pick off affected leaves and apply pre-ventive insecticide.
Tomato moth	Green caterpillar eats into stems, leaves and fruit.	Not a frequent pest, appear-ing spas-modically.	Apply insecticide.
Woodlice	Attack base of stems.	Infestation arises from	Insecticidal dusts or burning congregating

Trouble	Symptoms	Cause	Control
Mice	Eat seed or nibble young plants.	various sources. Infestation arises from various sources.	insects. Trap or bait timeously.
Wireworms	Enter main stem and tunnel up centre.	Infestation arises from recent pasture land.	Insecticidal dusts or soil sterilisation.
Millipedes	Affect roots, causing wilting.		Insecticidal drenches.

MINERAL DISORDERS

Mineral disorders can arise very frequently under the intensive cultural conditions practised in greenhouses, and perhaps more especially where a soil-less medium is used. A sensible approach to tomato nutrition can help in the avoidance of problems, but with the best will in the world disorders can occur. Conversely it is all too easy to become supersensitive to mineral disorders and blame them for every growth irregularity, when in fact it could be a pest, disease, water or temperature problem. The accurate diagnosis of mineral deficiency is not simple and the table opposite deals with the more general and easily recognisable symptoms. (See Chapter 4 for details of major element deficiency and excess.)

Mineral	Deficiency	Excess	Corrective treatment (if any)
Iron	Yellowing or whitening of whole leaves, except veins which remain green. Excess calcium in soil can result in iron 'shut-off'. Frequently seen in young rapidly growing plants as a mottling.	Difficult to detect visually and may seldom arise as excess iron tends to be combined with calcium to form insoluble calcium salts.	Check pH of soil. Apply chelated iron preferably when symptoms occur in the initial stages.
Magnesium	Interveinal yellowing on lower leaves first, gradually moving up plant. Orange areas develop. Worse on certain soils where drainage is free and where plants receive full sun. Excess potash application will frequently induce magnesium deficiency.	Not usually a problem apart from effect on salt concentrations.	Apply FYM to soil, especially if light, during soil preparation. Ensure that magnesium sulphate is applied before planting in future (see chart on page 123). Spray with magnesium sulphate (Epsom salt) 2% solution, 2lb in 10gal plus spreading agent, on several occasions when plants are young. Magnesium limestone can be applied instead of ground limestone.
Manganese	Mottling of leaves, often encountered in young plants.	Occurs in acid soils, especially when steam sterilisation is carried out frequently. Black blotches in stems with curled leaves.	Check pH of soil. Apply 2% manganese sulphate solution if deficiency is very bad.
Boron	Brittle leaves (with purple colouration)	Should seldom be a	Add boron at 1oz per gal of stock solution

Mineral	Deficiency	Excess	Corrective treatment (if any)
	and scorched leaf margins. Fruit may show necrosis of cell layer beneath skin.	problem.	(see page 163), or apply boron very *lightly* to soil at 1oz per 20sq yd and wash in.
Calcium	Whitening of growing ing head of plant, 'black bottoms' on fruit. Deficiency will arise in soil-less media due to calcium im-mobilisation by high ammonia/potash ratio (high salt concentra-tion).	May result in iron deficiency.	Apply ground lime-stone and wash in, or use calcium sulphate (1oz. to 1gal). Flush out media with plain water if soil-less system to reduce salt concentration.

PHYSIOLOGICAL TROUBLES

These troubles (opposite) do not appear to be directly related to a specific pest, disease or mineral disorder and so are called 'physiological'. Excess or insufficient water, the effect of temperature extremes, lack of air in the growing medium because of compaction, are frequent causes. The point must be made, however, that it is wrong to try and think in too compartmentalised a manner, as physiological disorders may have pest, disease or nutritional implications.

Symptoms	Main cause	Corrective treatment (if any)
Blotchy ripening Fruit showing light patches which fail to ripen.	Irregular watering and feeding resulting in variable salt content in soil.	Avoid irregular watering, feeding and temperature as far as this is practical. Some varieties are worse than others, especially those in the 'vigorous' categories.
Greenback The shoulder of the fruit remains green.	Common with green-back varieties such as Ailsa Craig and its off-spring. Moneymaker types were developed to avoid this disorder. Excess sunlight, lack of potash, or too hard defoliation can induce greenback.	Avoid over-defoliation. Shade in extreme instances when very hot weather persists. Step up potash application.
Blossom end rot or 'black bottoms'. The bottom end of fruit develops an area of cell necrosis which turns brown or black.	Due to high salt concentration in growing medium, also calcium deficiency. Lack of water also critical at early stages of growth.	Regular water applications. Flush out salts and start again with a balanced feeding programme.
Bronzing Dead layer of cells immediately below skin.	Excessively high daytime temperatures. Could also be due to virus or to boron deficiency; some confusion still exists as to the exact cause.	Avoid high daytime temperatures and check for other troubles. The trouble seldom persists for more than a few trusses.
Fruit splitting Fruit splits or cracks.	Due to irregular water uptake or occasionally to varying temperatures. A frequent trouble in cold houses when fruit has	Even up temperatures as far as possible, shading if necessary to avoid excess daytime temperatures. Go on to a much more

Symptoms	Main cause	Corrective treatment (if any)
	been a long time developing and has grown a tough skin.	regular watering pattern.
Oedema Nodules appear as bumps or blotches on stems. Transpiration cannot take place rapidly enough through leaves.	Due to continuous excess humidity or the application of oil-containing sprays. Common on container-growing systems where excess water lies about between troughs or containers.	Reduce humidity by getting rid of excess water. Adequate ventilation, especially at night.
Leaf curling Leaves develop excessive curls upward, especially on older leaves.	Due to large variation in temperatures between day and night (the plant being unable to cope with surplus carbohydrates).	Adjust day/night temperature differential.
Silvering Foliage turns light in colour on a portion of the plant; half leaves may be affected.	This is thought to be a genetical disorder associated with tissue layers and may correct itself or persist.	

FLOWER   DISORDERS

Dry set Pollen goes dry on stigma preventing fertilisation, leaving 'pin-head' fruits.	High temperatures, dry air, energy lag in plant resulting in 'check' period when infertile pollen may be produced. Virus disease can also cause poor setting.	Attempt to improve control of environment and check for virus.
Flower abortion Fertilisation occurs but abortion also occurs, ie the process of fruit	Considerable research is still being carried out on this phenomenon. It is thought to be related to	Propagation regime should be controlled as far as this is possible. The use of artificial light, either supple-

Symptoms	Main cause	Corrective treatment (if any)
development subsequent to fertilisation is arrested. Generally in first or second truss of fruit and only 'chat' fruits develop.	light input (light intensity) or day/night temperature relationship, which produces flowers bearing pollen not fully viable. Vegetative development of young plants is obviously a vital factor and this has been discussed in earlier chapters.	mentary or in a growing room, where there is some standardisation of conditions under which early flower trusses are laid down. (This of course occurs at a *very* early stage and the first and possibly the second trusses are initiated during the light-treatment period.)
'Missing' flowers Truss forms but flowers either do not appear or only open partially.	Due to excess supplies of nitrogen and connected with low salt content of soil. Low night/high day temperatures are contributory.	Apply potash to harden growth and adjust salt concentration in soil. Balance up day and night temperatures.
Flower drop Flowers drop off at 'knuckle'.	Due to lack of water at roots, dry atmosphere or high salt concentration in soil.	Check water and feed applications and improve environmental control.

IMPORTANT NOTE

Tomatoes are highly susceptible to contamination from weed-killing hormone materials which gain entry through vents etc. Leaves go nettle-like and fruit becomes pear-shaped with a 'carbolic' taste. Metham-sodium sterilants may cause acute leaf distortion in young plants.

13

Sterilisation of Soil

Any system of monoculture involving successive short-term crops gives rise to a variety of disorders of which nutritional imbalance is one. Each crop makes its special demands on available major and trace nutrients. In addition plant toxins, the acids excreted by all growing plants, accumulate and may influence adversely the micro-organism population of the soil or growing medium (soil-less media, although relatively sterile initially, rapidly becomes invaded by micro-organisms), which in turn has nutritional repercussions. Antagonism and competition can also develop between the micro-organisms, beneficial groups frequently being repressed by non-beneficial ones. The build-up of fungal diseases and harmful pests specific to any crop, in this case tomatoes, is of course of paramount importance. Matters are not helped by the widespread use of chemical fertilisers, which further give rise to complications in the build-up of toxic residues.

The simple term for such a motley collection of problems is 'soil sickness' and it is certain that organic principles of husbandry could do much to prevent it, always provided of course that one is prepared to wait until troubles have been overcome—but this is not always practicable, especially in commercial circles.

Soil sterilisation, or more accurately pasteurisation, either by heat or chemical means, endeavours to rectify these ills and leave the beneficial elements to function. Pasteurisation is, technically speaking, different from complete sterilisation which renders a medium sterile, not a desirable state of affairs for soil-containing medium, but the term soil sterilisation is nevertheless commonly used to mean pasteurisation. Weeds in both vegetative and seed form should be killed by effective sterilisation. Sterilised soil is not immunised in any way, and indeed is perhaps more prone to re-infection by troubles than one which is not sterilised.

Sterilisation by the use of heat is, however, far from being an exact technique and has definite limitations unless adequate temperatures are achieved for a suitable period of time and all parts of the growing medium are equally sterilised—simple in theory but difficult in practice. This is especially true in the control of virus diseases and potato eelworm. Ammonia release is also a problem in soil-based media, as the main groups of bacteria left unscathed are thick-walled ammonia and nitrogen-forming types.

The many problems of sterilisation, coupled with the shortage of reliable and consistent supplies of soil, are the main reasons why such widespread use has been made in recent years of soil-less media, where the basic ingredients do not normally require sterilisation if fresh supplies are used on a regular basis.

The last few years have seen considerable changes in sterilisation techniques; the following sections give an account of modern methods (see Ministry of Agriculture Bulletin No 22 for full technical details). It must be appreciated that certain methods are obviously impossible in amateur circles, both in respect of the necessary equipment and because of the undesirability of using certain chemicals—legislation existing in the latter case.

HEAT STERILISATION

Several methods of heat sterilisation are practised, with the basic object of bringing the soil or growing medium uniformly to a temperature sufficiently high and for a long enough period to kill off harmful organisms, while leaving beneficial types unharmed. The period of heat should not be prolonged enough to upset the nutritional balance. The temperature of 160–170° F (71–77° C) is generally accepted as being effective for most harmful organisms, although virus-infected debris and possibly eelworm cysts may remain unscathed, necessitating higher temperatures of 180–190° F (82–87° C) as confirmed by a soil thermometer (higher temperatures are frequently advised).

The Hoddesdon Method

This method was developed at the Cheshunt Research Station (since closed). It involves leading pressurised steam into perforated pipes placed 10–12in below the soil, which should preferably be moist and friable. A PVC sheet is placed on top of the soil and steam applied for 15–20 minutes or longer until the *top* of the soil has reached 180–190° F (82–87° C). Spikes or harrow grids are also employed, the latter being pulled through the soil by an automatic winch.

Sheet Steaming

Developed originally in the Netherlands and modified by H. B. Wright of Cottingham, this process involves leading the pressurised steam into either perforated hose or a 'coffin'-type box with outlet holes under layers of heavy grade PVC sheeting secured at the edges by sandbags or other means. The sheet billows up and the steam condenses on the soil surface, the heat travelling downwards in bands. Penetration of heat depends both on soil type and method of cultivation, the best penetration being where soil is of a porous type, or rendered so by rough digging or some other method which does not make the soil too fine. Soil temperature must be accurately measured by side or oblique insertion of a soil thermometer. It can take several hours for the steam to penetrate deep enough, and in an effort to improve penetration stronger PVC sheeting is often used or nylon net is put over the PVC. Methods involving the using of air and steam to bring the temperatures down sufficiently to avoid over-sterilisation of the soil surface are also practised. Sheet steaming has found favour with growers of lettuce, chrysanthemum and short-term tomato crops, although results have been excellent in many instances for long-term tomato crops, much depending again on the presence or otherwise of deep-seated troubles. The main advantage of sheet steaming is the saving in labour.

Low-pressure Steaming and Other Methods

Few smaller growers or amateur gardeners have facilities for producing pressurised steam in sufficient quantity for effective sterilisation, although small quantities of soil for potting purposes or the preparation of composts can be steamed with a small electric steam steriliser. Low-pressure steaming is, however, an acceptable proposition, whereby water is boiled below a perforated tray which contains the soil, a PVC sheet again being used to cover the soil. Low-pressure sterilisers can be built to the John Innes specification or bought in mobile wheeled form. Low-pressure sterilisation can be slow, according to soil type and moisture content, it being highly important to check the temperature achieved on the soil surface.

Drenching shallow layers of soil with boiling water and covering immediately with *clean* sacks has much to commend it, as also has the suspension of small quantities of soil in a sack over boiling water. Soil so treated can be reasonably well sterilised.

Dry Heat Sterilisation

This can take several forms, the most basic being the use of a fire over which a sheet of corrugated iron or a sheet of metal is then placed and the soil spread in shallow layers above the main source of the heat. It helps if the soil is fairly wet, and care must be taken not to over-sterilise the soil or burn off the organic matter content, which will completely destroy soil structure as well as rendering it sterile. A flame gun can also be used to the same effect for limited quantities of soil spread in a shallow layer on a clean surface.

Rotary drum sterilisers are widely used on commercial holdings and by local authorities. The soil is put in at one end of the drum which, by rotating, allows the soil to fall several times through a flame. The angle of the drum determines the rate at which the soil is ejected out of the other end of the drum. Care must be taken that the soil is sufficiently dry, otherwise it can come through improperly sterilised, frequently 'balling' in the process. It is highly important to check the temperature of the ejected soil to ensure that it comes out at a uniform 160–170° F (71–77° C) or higher if virus or eelworm control is sought.

Electric sterilisers operating on the panel element system are highly efficient and available in several sizes.

CHEMICAL STERILISATION

Various chemicals are used for sterilisation and for economic reasons have become very popular in commercial spheres. Chemicals are also useful for the amateur gardener, but great care must be taken to ensure that they are properly used. The main drawback to chemical sterilisation is that the fumes are toxic to young plants, which raises problems in 'mixed' greenhouses.

To be effective, chemicals must give off gas of sufficient toxicity and concentration to be capable of destroying pests, diseases and weeds, and yet leave the beneficial organisms unharmed. The efficiency of chemicals varies greatly, not only according to their precise chemical content, but according to the condition of the soil, its moisture content and, perhaps more important, prevailing temperature. Even distribution throughout the soil being sterilised is also highly important.

Formaldehyde

This has a good effect on mild fungal diseases, but is relatively ineffective against many pests. It is used at a strength of 1 part formalin 38–40 per cent to 49 parts water (approximately 1 pint in 6 gallons) and the soil is drenched thoroughly, about 5 gallons per square yard being required (to 9in depth). The soil is ready for use in 20–40 days according to temperature, which must be sufficiently high otherwise the formaldehyde will merely polymerise.

Phenol and other chemicals

Various phenols are available, including cresylic acid, and should be used according to directions, but broadly in a similar way to formaldehyde, the soil being ready for use in 18–20 days. Action on the soil is largely insectidal, but there is also some fungicidal action. Usually about 5–6 gallons of the prepared solution is required per square yard.

Other chemicals used commercially are D-D (principally on nematodes) and carbon bisulphide, both of these being injected into the soil, dithane, chloropicrin, methyl bromide, and metham-sodium. The metham-sodium sterilants are widely used commercially. Available as Campbell's Metham Sodium, Sistan, Solnam, Trimaton Unifume, Vapam, Vitafume (all liquids) and Dazomet which is a powdered or prilled form.* The metham-sodiums have a wide control spectrum. These chemicals must be used strictly according to directions, which generally state that the soil should be cultivated, the sterilants applied and, with the powdered or prilled form, rotovated in before 'sealing' with water. For them to be effective the temperature of the soil must be above 49–50° F (9·5° C). After a few weeks the soil should be rotary cultivated or forked on several occasions to release the fumes and prevent damage to plants. Heaps of potting soil can of course be treated in a similar manner. Methyl bromide has found increasing favour in recent years, due to its excellent action on pests and diseases, and also to its rapid dissipation. *It must be applied by a contractor.*

Sterilisation Chart overleaf

* Methyl isothiocyanate (Basamid), a slightly different formulation.

STERILISATION CHART

Material	Root eel-worm	Root rots	Vert wilts	Virus (TMV)	Damp-ing-off disease	Weeds	Days between planting and treatment
Steam (including properly applied dry heat and hot water)—180°F (82°C)	C	C	C	C (good at 194°F (90°C))	C	C	7–14
Metham-sodium* (results variable)	C	C	P	X	C	C	40
Formaldehyde	X	P	P (good at 60°F 16°C)	X	P	X	20–40
Dithane	X	P	X	X	P	X	3–14
Cresylic acid	P	X	P	X	P	X	18–20
D-D (by injection)	C	X	X	X	X	X	40
Carbon bi-sulphide (by injection)	C	X	X	X	X	X	7
Methyl bromide†	C	C	P	X	C	C	4
Chloropicrin†	X	C	C	X	C	X	20

Key: C = control
P = partial control
X = no control
* = only certain forms may be available to amateurs
† = not available to amateurs

Note: seed boxes, pots, etc, should also be sterilised, although with the increasing use of plastics, washing out with hot water containing a little detergent is usually sufficient.

14

Tomato Varieties

The range of tomato varieties offered today in Britain is extensive and if one takes into account the different types of tomatoes, including the large fruited multi-celled form popular in the USA, varieties run into many thousands. Some rationalisation is, I feel, long overdue, as there are many varieties so similar in nature that they ought not to be named differently. A further problem is that there are many different strains of any one variety and this means in practical terms that, say, Ailsa Craig from one source may be different from Ailsa Craig from another source. This is due to a gradual change in breeding lines, and the differing criteria for selection exercised by seed raisers. The choice of any particular variety, moreover, does not automatically ensure cropping results to a set standard, as environmental conditions are bound to vary enormously, and these affect variety performances.

GENERAL CLASSIFICATION OF VARIETIES

Broadly speaking, tomatoes can be classified in the different ways listed below.

F1 Hybrids

These have been available for many years and are the first seed generation following a controlled cross between selected male and female parents. Seed must therefore be produced annually, otherwise there would be segregation into respective parental groups. The virtue of the F1 generation is that the inherent characteristics of the parents can easily be combined, and this is of special significance with regard to vigour, disease resistance and fruit form.

Vigour

There are broad groups, although somewhat nebulous, into which varieties are categorised.

Compact varieties. Where plants have short internodes, both height and breadth are limited, with root systems of similar habit. Compact varieties are useful where very strong growth is likely to be a problem (following the steaming of an organic soil) or where height is restricted.

Tall spreading or vigorous varieties. Typified by Ailsa Craig, these varieties make big roots and vigorous leafy growth, usually (but not always) with long internodes. There can be a problem when growing vigorous varieties in a very rich soil.

Intermediate varieties. These are not now generally categorised in seed catalogues and their placing in the variety list which follows cannot therefore be too definite.

Fruiting Potential

Varieties obviously vary in the weight of fruit they can produce, although much will nevertheless depend on growing treatment, number of plants in a given area, and many other factors. Varieties with short internodes will produce more fruit per foot of stems than plants with long internodes, but on the other hand the vigour of any variety can lead to greater total fruiting potential.

Much more important is the average number of fruits per truss, and whether single, double or treble trusses are formed. Highly important also is the ease of setting, this being influenced by the production of viable pollen.

Earliness of Cropping

This is a very important feature for the commercial grower, some varieties being able to stand 'forcing' while others, if given the same treatment, will go vegetative. The ability of a variety to produce early fruit is tied up with the number of days from flowering to fruit-ripening and the amount of heat and light needed to accomplish this.

Disease Resistance

This is of considerable importance and includes resistance to the various strains of cladosporium, resistance *or* tolerance to TMV, resistance to various wilts and, in the case of root stocks, to corky root and root knot eelworm. Other resistances are constantly being developed by breeders, both in Britain, Holland and elsewhere (see notes on page 186).

Fruit Size, Form and Quality

Varieties also differ in that they give greenback, greenback free, or semi-greenback fruits, large or small fruits, flat or round shaped, and fruit of contrasting colours.

MAKING A CHOICE OF VARIETY

In making a choice of a specific variety a gardener or grower will not only consider local experience related to prevailing climate and soil, but will endeavour to obtain resistance to specific disorders. The variety lists that follow do not claim to be exhaustive. New varieties are constantly being produced and readers are advised to make constant perusal of seed catalogues. (See Acknowledgements, page 232, for the names of seed firms which have assisted in the compilation of these lists.)

TOMATO VARIETY LIST

This list includes most of the varieties that appear in the major British seed-house catalogues at the time of writing, but, because of lack of information on certain varieties, it cannot be fully exhaustive or absolutely accurate. Variety performance also differs considerably according to environmental conditions and no general statement made here can be taken as true of performance in all circumstances. Some of the brief remarks in the comments column are based on personal experience or on information supplied by Mr L. A. Darby of the Glasshouse Crops Research Institute and Mr F. Holl of Asmer Seeds, but in many cases they are from seed catalogue accounts and must not be taken as statements of fact following detailed trials.

The use of the asterisk * indicates reasonably certain expression of the stated characteristic.

The British seed trade fully appreciates the need for a review of tomato varieties; availability therefore is inevitably subject to constant change.

Variety	Early	Cropping pattern (Heavy = total crop; sustained = cropping well over period)	Tall	Inter-mediate	Com-pact
Fɪ HYBRIDS					
Amberley Cross (GCRI)	*		*		
Ailresist	*		*		
Arasta			*		
Asix Cross (Bruinsma)		Heavy	*		
Bonset				*	
Clavito				*	
Cudlow Cross (GCRI)		Heavy	*		
Delkro (Pannevis)		Sustained	*		
Emgro	*			*	
Eurocross A (Bruinsma)				*	
Eurocross B (Bruinsma)				*	
Eurocross BB (Bruinsma)	*			*	
Extase (Enzazaden)		Heavy	*		
Findon Cross (GCRI)	*				*
Florissant (Bruinsma)			*		
Fontwell Cross (GCRI)	*		*		
Globeset			*		
Growers Pride	*		*		
Ijsselcross (Bruinsma)			*		
Jet (Enzazaden)	*	Heavy		*	
Kingley Cross (GCRI)		Sustained			*
Kirdford Cross (GCRI)		Sustained			*
Lavant Cross (GCRI)	*		*		
Maascross (Bruinsma)	*		*		
MG (Pannevis)			*		
MM (Pannevis) incl different strains such as Super, Nova, Mɪ3, Extra					
Moneyglobe	*			*	

Wilt resistance F = Fusarium V = Verticillium	TMV Resistance (R); tolerance (T)	Clado-sporium resistance (A or B)	Green-back free	Out-door	Non-heated or heated houses	Comments
		A & B	*			Good quality fruit
		A & B				Good for poor-light areas
		Not specified	*			Good taste and shape
		A & B	Semi			Early planting
V		A & B	*			Large rough fruits
	T	A & B	Semi			Autumn cropping
F		A & B	*			
		A & B	*			
		A & B				
		A	*			Fairly early
		A	*			For soils of lower fertility
		A	*			Large fruits, early planting
		A & B	Semi			Good shape, free setting
		A & B	*			Introduced 1971
		A				Good quality fruit
F		A & B	*			Introduced 1971
		A & B	*			
		A				Vigorous grower
		A & B	*			Larger fruit than Maascross
	*	A & B	*			
		A & B	*			Labour-saving variety to grow
	R	A & B	*			Introduced 1971
V		A & B	*			Introduced 1971
		A	*			Good quality
		A	*			Ailsa Craig type
		A or A & B	*			Excellent range of varieties of Money-maker type
		A	*			Sets freely

Variety	Early	Cropping pattern (Heavy = total crop; sustained = cropping well over period)	Tall	Inter-mediate	Com-pact
Gannet (GCRI)	*		*		
Grenadier (GCRI)	*		*		
Odine			*		
Pagham Cross (GCRI)		Sustained	*		
Panagro (Pannevis)	*	Heavy	*		
Panase (Pannevis)			*		
Plusresist	*			*	
Red Ensign (Hurst)		Heavy	*		
Rijncross (Bruinsma)			*		
Ronald 'M'			*		
Selsey Cross (GCRI)		Heavy	*		
Seriva			*		
Sonata (Van den Berg)					
Single Cross (Bruinsma)	*	Heavy	*		
Supercross (Bruinsma)			*		
Surprise C.70 Enzazaden				*	
Syston Cross (Harrisons)			*		
Top Cross (Bruinsma)		Heavy and sustained	*		
Virocross (Bruinsma)		Heavy	*		
Ware Cross	*		*		
Witham Cross	*		*		

STRAIGHT VARIETIES (vigour groups all rather indefinite)

Variety	Early		Tall	Inter-mediate	Com-pact
Ailsa Craig	*		*		
Alicante (Suttons)	*		*		
All Round			*		

Wilt resistance F = Fusarium V = Verticillium	TMV Resistance (R); tolerance (T)	Clado-sporium resistance (A or B)	Green-back free	Out-door	Non-heated or heated houses	Comments
		A & B	*			
		A & B	*			
	*	A & B				
	R	A & B	*			Introduced 1971
		A & B	Semi			For less fertile soils
		A & B	Semi			Strong top growth
		A & B	Semi		*	Firm round fruits
		A			*	Fast growing
		A & B	Semi	*		
		Not specified				Specially bred for cold growing
		A & B	*			Reliable
	T	Not specified				Has performed well in trials
						Round fruit
	T	A & B	*			Fruit like Money-maker
		A & B	*			
		A				Free-setting, good quality
		A			*	Very good yields recorded
	T	A & B	Semi			Short jointed / Reliable cropper
		A & B	*			Has performed well in trials
						Possibly the best known tomato ever. Many strains exist
		Not specified	*			Performed well in trials
		Not specified	*			A Dutch variety

Variety	Early	Cropping pattern (Heavy = total crop; sustained = cropping well over period)	Tall	Inter-mediate	Com-pact
Antimould A (Blackpool strain)			*		
A1 (Webbs)		Heavy	*		
Best of All		Heavy		*	
Craigella (GCRI)	*		*		
Discovery		Heavy		*	
Early Beauty (Webbs)			*		
ES 1			*		
Exhibition		Heavy	*		
First in the Field			*		
Fortuna			*		
Golden Dawn			*		
Golden Perfection					
Harbinger	*		*		
Hundredfold			*		
Jubilee Improved			*		
Leader (Suttons)	*				
Market King			*		
Marton Special			*		
Mid-day Sun					
Minimonk (GCRI)					*
Moneymaker				*	
Pipo				*	
Open Air			*		
Outdoor Girl					
Potentate				*	
Puck					*
Scarlet Emperor			*		

Wilt resistance F = Fusarium V = Verticillium	TMV Resistance (R); tolerance (T)	Cladosporium resistance (A or B)	Greenback free	Outdoor	Non-heated or heated houses	Comments
		A			*	Good for late planting
			*			Fruit of good colour
						Reported to be huge cropper
			*			Good eating quality
			*			Variety with excellent reputation
						Excellent taste
			*			Rough fruit
			*	*		
			*	*		Potato leaved
						Yellow even-sized fruit
						Yellow fruit
				*	*	Good quality
				*	*	Good quality
		Not specified				
				*	*	Smooth fruit
				*	*	
				*	*	Much grown at Blackpool
			*			
			*			Good for bench growing
			*			
						Self-stopping at 4 trusses
				*		
				*		Large fruit of poor quality
				*		Ideal for cloches

Variety	Early	Cropping pattern (Heavy = total crop; sustained = cropping well over period)	Tall	Inter-mediate	Com-pact
Sunrise			*		
Stonors MP			*		
Stonors Dwarf Gem					*
The Amateur					*
Tigerella (GCRI)	*		*		
Tiger Tom (GCRI)	*				
Tangella (GCRI)	*		*		
Victor & Victor 70			*		

Note. Pelleted seeds of certain varieties can now be obtained to facilitate space sowing.

Wilt resistance F = Fusarium V=Verticillium	TMV Resistance (R); tolerance (T)	Clado-sporium resistance (A or B)	Green-back free	Out-door	Non-heated or heated houses	Comments
		Not specified				An old but still popular variety
				*		Small fruits
				*		Bush variety, no staking
			*			Golden stripes on red fruit
			*			Striped fruit
			*			Tangerine anthers and fruit

Notes on Variety List overleaf

NOTES ON THE VARIETY LIST

Cladosporium resistance is stated as A or B or a combination of both. Strain C is not yet a problem in Britain. *Virus resistance and tolerance* is stated. Obviously one should not grow a tolerant variety alongside a non-tolerant one.

Availability

All the varieties listed are currently available through British seed houses in 1971, and amateur gardeners having difficulty in procuring any variety are advised to approach one of the major seed firms listed in the Acknowledgements. Varieties raised by the GCRI (The Glasshouse Crops Research Institute, Littlehampton, Sussex) and by Dutch growers are all offered through the seed trade.

Virus Testing and Heat Treatment

Virus testing of seed refers to the technique suggested by Dr L. Broadbent of Bath, in which a small portion of the selected seed is used as an inoculation for injecting indicator plants. It is carried out on a 'spot check' basis. Heat treatment of seed refers to the treatment of the seed to free it, as far as possible, from seed-borne virus. Germination may be delayed by this treatment. Whether or not seed is tested and treated, or both, is usually stated in seed catalogues.

Saving Seed

For those saving seed of STRAIGHT VARIETIES ONLY, the following is the procedure. Select two or three plants of good habit and allow the fruit on the selected truss (usually the third or fourth) to become fully ripe. Cut the fruit in two and scrape out the seed into a glass, plastic or glazed earthenware container. Alternatively the fruit can be squashed through a ¼in sieve into the container to partly separate the pulp and skin from the seed. The seed is left in the container for 5 to 6 days in a warm place so that it will ferment and so allow the seed to completely separate itself from the skin and pulp. After this it can be put in a fine sieve and washed carefully with a strong jet of water before being placed on a clean sheet of glass to dry.

Alternatively the seed slurry is well mixed with one quarter of its volume of concentrated hydrochloric acid (35–38 per cent), and left for half an hour before being washed clean. This is a method much used by seed raisers to try and ensure freedom from virus-infected debris on the seed coat. The seed is again dried on a clean sheet of glass. After drying, it is then stored in packets in a dry place, being marked with variety and date.

Seed of compact habit varieties should not be cleaned by this method as it impairs their germination. Use instead an equal volume of 10 per cent sodium carbonate and leave for 24 hours before washing clean.

15

Outdoor Tomatoes

The success of tomatoes grown entirely out of doors, apart from the propagation period, is related entirely to weather pattern as influenced by latitude, exposure, and type of season. In many years it is impossible to produce ripe fruit outdoors in many areas of Britain. Often protection must be provided by means of cloches or polythene shelters. However, the availability of varieties of prostrate growing habit and those with the ability to grow and produce ripe fruit under lower temperature regimes, has given a boost to the growing of tomatoes out of doors.

Favoured areas of Britain such as the south-coast region, including the south-east corner round to Essex, are perhaps the ideal areas for outdoor culture, and success has also been achieved by gardeners in other eastern counties such as Norfolk. The Channel Islands, and Jersey in particular, have long grown outdoor tomatoes commercially, although in recent years even Jersey has tended, like its sister island Guernsey, to move more towards culture of tomatoes under glass, because of the unpredictable nature of the outdoor crop.

PRE-PLANTING PROCEDURES

Site Selection

Within the area available a site should be selected which offers protection from wind and receives the maximum amount of sun. Should a south-facing border in front of a tall wall be available, this is ideal. Alternatively, select the best open site there is, taking into account such factors as wind. The use of portable shelter materials (see page 21) can often improve facilities enormously.

Soil Preparation

The instructions in each chapter relating to the nutrition of tomatoes, while relevant, are not all completely applicable. Outdoor tomatoes are of remarkably short-term nature and therefore nutrient uptake is much less than for a glasshouse crop. Conversely however there is not so much control of water application and there could be excessive leaching out of nutrients during wet spells of weather. Soil texture and drainage should be good, and well-rotted FYM or peat should be applied in sufficient quantities to bring about the necessary improvement, coupled with deep digging to improve drainage. Quantities of FYM can be in the region of one barrowload per 6–8sq yd, and peat in almost unlimited quantities, provided the necessary amount of lime is used to offset its acidity (see page 119).

Alternatively special raised beds, can be prepared, using compost (John Innes Potting No 2 type) or a soil-less equivalent in ridges, roughly on the basis of half a bushel per plant—on top of polythene, if necessary, where pests or diseases are likely to be a problem. Growing in polythene-lined trenches filled with soil-less media can also be worth considering, provided there are drainage facilities.

Tomatoes although a short-term crop out of doors, are still susceptible to various maladies, potato eelworm being a particularly likely pest in gardens where potatoes have been grown. Wireworms can also be troublesome. While chemical sterilisation of soil out of doors is perfectly possible, and is practised commercially in Jersey, it is seldom practical in the average garden. A measure of crop rotation should, however, be possible. But obviously, if there is difficulty about this the alternative cultural systems referred to throughout the book can be considered, on the lines described for indoor systems.

Application of Lime

The adjustment of the pH to around the 6·5 level should be carried out, avoiding direct contact of FYM and lime (see page 120) and applying the lime in time to allow the pH figure to rise before planting. In areas of calcium soil no lime will of course be needed and there is much virtue in adding copious amounts of peat to reduce the pH sufficiently to avoid iron-deficiency problems.

Application of Base Feeds and Final Preparation of the Soil

Where special composts are not being used, a base feed should be applied at 6–8oz per square yard, selecting one of medium potash content (see pages 64 and 123). Base feeds should be evenly applied and well raked in, bringing the soil to a final state for planting, that is relatively firm, yet with a good tilth on the surface.

PLANTING

Planting distances for outdoor tomatoes are in the region of 15–18in apart in rows 30–36in apart, running the rows north-south if there are several, or east-west where only a few rows are planted, this being especially the case in south-facing borders backed by a south-facing wall.

Propagation and Time of Planting

Reference to the propagation table on page 91 will show that as the year progresses less time is taken to propagate tomatoes, owing to the better light pattern. Planting of outdoor tomatoes is unlikely to be considered until around late May in the south (perhaps a little earlier in the Channel Islands) and early June in the north. Allowing the 4–6 weeks necessary for propagation, it can be seen that it is necessary to sow the seed around mid-April. Obviously if little heat is available, as is often the case where outdoor tomatoes are grown anyway, there will be a growth rate commensurate with prevailing temperatures.

Plants can be grown in 4¼in pots or in large soil blocks and brought to first truss flowering stage in a cool greenhouse or raised open cold frames before planting out firmly with the soil ball slightly below soil level, making sure the plants are fully acclimatised to outside conditions.

Establishment of Plants

Provided the plants are suitably hardened off and are not pot bound (they may have been kept too long in their pots waiting for suitable planting weather), they should establish fairly rapidly, this being assisted by a light watering or spraying with clean water in hot weather. Stake immediately after planting (except for bush varieties—see page 215) with 4½ft canes. Some temporary shelter, especially from wind, can help establishment. Plants invariably assume a darker green colouration outdoors than plants growing in a greenhouse, and there can often be a tendency towards hardness of foliage, coupled with an upward curling of leaves, especially when cool nights follow warm days. This is because the carbohydrates manufactured during the day in the plant's leaves are not dissipated at night under cool conditions, as they are in a warm greenhouse.

TRAINING AND GENERAL CULTURE

Plants are secured carefully to their 4½ft canes, using soft twine or paper/wire clips, a practice which must be continued regularly throughout the development of the plant. Side shooting should be practised as for indoor tomatoes (except for bush varieties) and the plants pinched out above the third to the fifth truss, according to the district and the exact situation involved. Obviously if growth is slow, as it can be in a cool summer, it may only be possible to ripen two or three trusses, remembering that once the fruit forms and increases in size, heat and warmth are still needed for ripening. Once a shine appears on the fruit, this is an indication that ripening is not far distant. On average four trusses should be possible.

Plants must never be allowed to dry out, and a mulch of straw or peat helps in this direction. Spraying overhead in very dry weather will help setting. Feeding should be carried out regularly every 10–14 days with a liquid feed (see page 163), it is generally the case that medium potash types are required for much of the season, rather than high nitrogen feeds which tend to lower fruit quality—although the precise nature of feeding can be adjusted according to the rate of growth.

Bush Tomatoes

Cultural pattern is much the same, except that plants are not supported with canes or side-shooted, merely being allowed to sprawl over the ground on a deep mulch of straw or low-set wire-netting. Modern breeding has produced bush tomatoes which are more productive at lower temperatures than 'standard' varieties.

Cloche Culture

Several different techniques can be adopted using conventional cloches or polythene tunnels. Where cloches are to be used merely to start the crop off, planting out is usually carried out after the cloches have been in position for some 10–14 days to allow the soil to warm up. Planting can generally be carried out 2–3 weeks earlier than planting outdoors, and obviously there are greater benefits to be derived in exposed areas. Cloches can be removed when the plants have developed sufficiently, and the plants then staked and trained as for outdoor culture. In early September a mulch of straw can be laid down and the plants removed from their stakes and laid down under the cloches to allow green fruit to ripen, which it can do reasonably well in a good autumn. Polythene tunnels can be used in a similar manner.

Alternatively, if the cloches are large enough, and there are several elevated types of cloches available, the plants are grown under them throughout the full season. Humidity under the low polythene cover is a factor to be taken into account, and ventilation is important, as is regular watering despite any lateral movement of soil moisture. Training systems under cloches are varied, but generally involve running a wire horizontally on stakes 4in below the top of the cloche, tying the plant to this. The number of trusses which can be grown under cloches varies enormously, and results are usually commensurate with the degree of care exercised. Some defoliation will definitely be required, as for greenhouse culture, but this should not be carried out to excess.

Bush varieties lend themselves admirably to full season culture under cloches, the main point to watch being regular watering. It helps if trusses are supported on a low-set wire, or alternatively wire-netting kept a few inches above ground level.

Shading with a proprietary shading material or lime and water may be necessary in extremely hot weather.

Pests and Diseases

These follow a similar pattern to greenhouse tomatoes—potato blight being the main outdoor plague.

Varieties

See general list of varieties (Chapter 14).

Pest and Disease Control in Tomatoes

List of chemicals from the Agricultural Chemicals Approval Scheme booklet *Approved Products for Farmers and Growers—1975* revised annually). Consult this booklet for trade names of chemicals listed below.

Aphids: BHC, demeton-S-methyl, derris, diazinon, dichlorvos, dimethoate, formothion, malathion, nicotine, oxydemeton-methyl, parathion, tecnazene/BHC, pirimicarb, propoxur

Blight: copper, maneb, zineb

Botrytis: benomyl, dicloran, tecnazene, tecnazene/BHC

Brown root rot ('Corky disease') and root rots generally: chloropicrin/dichloropropane-dichloropropene/methyl isothiocyanate, dazomet, metham-sodium, nabam, nabam/zinc sulphate, zineb

Leafhoppers: BHC, DDT, DDT/BHC, derris, malathion

Leafminer: BHC, diazinon, nicotine, parathion, tecnazene/BHC, DDT/BHC

Leaf mould (cladosporium): benomyl, copper, maneb, nabam/zinc sulphate, zineb

Potato cyst eelworm: chloropicrin/dichloropropane-dichloropropene/methyl isothiocyanate, dazomet, dichloropropane-dichloropropene, dichloropropene, metham-sodium

Red spider mite: azobenzene, demeton-S-methyl, derris, diazinon, dichlorvos, dicofol, dimethoate, formothion, oxydemeton-methyl, parathion, petroleum oil, tetradifon, tetradifon/malathion

Root knot eelworm: chloropicrin/dichloropropane-dichloropropene/methyl isothiocyanate, dichloropropane-dichloropropene, dichloropropene, parathion

Springtails: BHC, DDT/BHC, diazinon, parathion, tecnazene/BHC

Stem rot (didymella): captan

Symphylids: BHC, parathion

Thrips: BHC, BHC/tecnazene, DDT, DDT/BHC, derris, diazinon, dichlorvos, malathion, nicotine, tecnazene/BHC

Tomato moth: DDT, DDT/BHC, dichlorvos

White flies: BHC, BHC/tecnazene, DDT, DDT/BHC, dichlorvos, malathion, parathion

Woodlice: BHC, DDT, DDT/BHC, parathion, tecnazene/BHC

Notes on Tomato Grading

Important legislation was introduced in Britain during 1964—the Agriculture and Horticulture Act—which provides for 'extending the provision for assisting by way of the paying of grants the production and marketing of horticultural produce, and the imposing of requirements as to the grading of horticultural produce'. Commercial growers are now eligible for grants for the building of new and the replacement of old glasshouses, packing houses, grading machinery, modernisation of heating systems and many other specialised appliances and buildings. Statutory grading of specific horticultural crops followed this legislation and tomatoes, the subject of this book, were no exception. Statutory grading for this crop began on 13 May 1968, and supersedes the voluntary grading of the now defunct Tomato and Cucumber Marketing Board. It applies to tomatoes sold only through wholesale channels, which may be defined as wholesale markets, individual wholesalers' premises, grower/wholesalers, packing stations and depots for chain stores and supermarkets.

Grading scheme inspections are undertaken by Horticultural Marketing Inspectors (Marketing Officers in Scotland), who inspect consignments selected at random. The grades comprise requirements as to development, ripeness, colour, shape, cleanliness, progressive and non-progressive defects, blemish and sizing, and fall into four distinct classes: Extra Class and Classes I, II and III, in that order of merit.

Growers who have modernised their production units and are using improved production techniques have benefited from increased yields of crops which give a high percentage of tomatoes of good quality. Home produced tomatoes are now available fresher and of better appearance and quality than ever before. There are many who associate quality with flavour and are adamant that flavour has suffered in recent years. If, however, the correct variety is chosen for the appropriate soil and growing method, the characteristic flavour will emerge. Sceptics can then be told in answer to their remarks about flavour not being what it used to be that they must be referring to imported tomatoes, which are never as fresh, and indeed may be weeks old.

Metric and Other Conversion Factors

Definition	To convert	Into	Multiply by
Temperature rise	deg F	deg C	0·55
and fall	deg C	deg F	1·8
Length	in	mm	25·4
	mm	in	0·0394
	ft	metres (m)	0·3048
	m	ft	3·2808
Area	ft²	m²	0·0929
	m²	ft²	10·7639
Volume	ft³	m³	0·0283
	m³	ft³	35·3148
Mass	lb	kg	0·4536
	kg	lb	2·2046

BUILDING HEAT LOSS DATA

Thermal transmittance

U-value	Btu/ft² deg F	W/m² deg C	5·678
	W/m² deg C	Btu/ft² deg F	0·176

Useful data

20 fluid ounces = 1 pint
1 gallon of water weighs 10lb
Cubic capacity is calculated by multiplying length × breadth × average height

TEMPERATURE CONVERSION TABLE

°F	°C	°F	°C	°F	°C
86	30·0	60	15·6	34	1·1
85	29·4	59	15·0	33	0·6
84	28·9	58	14·4	32	0·0
83	28·3	57	13·9	31	—0·6
82	27·8	56	13·3	30	—1·1
81	27·2	55	12·8	29	—1·7
80	26·7	54	12·2	28	—2·2
79	26·1	53	11·7	27	—2·8
78	25·6	52	11·1	26	—3·3
77	25·0	51	10·6	25	—3·9
76	24·4	50	10·0	24	—4·4
75	23·9	49	9·4	23	—5·0
74	23·3	48	8·9	22	—5·6
73	22·8	47	8·3	21	—6·1
72	22·2	46	7·8	20	—6·7
71	21·7	45	7·2	19	—7·2
70	21·1	44	6·7	18	—7·8
69	20·6	43	6·1	17	—8·3
68	20·0	42	5·6	16	—8·9
67	19·4	41	5·0	15	—9·4
66	18·9	40	4·4	14	—10·0
65	18·3	39	3·9	13	—10·6
64	17·8	38	3·3	12	—11·1
63	17·2	37	2·8	11	—11·7
62	16·7	36	2·2	10	—12·2
61	16·1	35	1·7		

Formulae for conversion: Celsius to Fahrenheit $= (\,°C \times 1·8) + 32$
Fahrenheit to Celsius $= (\,°F - 32) \times ·55$

Acknowledgements

I am deeply grateful for the help given in the preparation of this book by the following friends and establishments:

G. C. S. Wilson, Chemistry Dept, West of Scotland Agricultural College

The late J. C. Raymond, Botany Dept, West of Scotland Agricultural College (for illustrations)

The Scottish Field, Buchanan St, Glasgow (for permission to use a number of photographs)

L. A. Darby, Glasshouse Crops Research Institute, Littlehampton

The late Duncan McArthur, West of Scotland Agricultural College

W. Godley, South of Scotland Electricity Board, Inverlair Avenue, Glasgow

The Electricity Council, Trafalgar Building, London, SW1

Acknowledgement is also made to the following firms for their assistance in the compilation of the variety lists on pages 210–17:

Asmer Seeds Ltd, 10 St James Street, Leicester

D. T. Brown & Co Ltd, Premier Seed Warehouse, Poulton-le-Fylde, Lancs

Finneys Toogoods Brydons, 94–104 Grainger Street, Newcastle upon Tyne

Sutton & Sons Ltd, Royal Seed Establishment, Reading, Berks

Nutting & Thoday, Long Stanton, Cambridge

Index

Page numbers in italic refer to plates

Adventitious roots, encouragement of, 178, 191

Aggregate, for ring culture, 125–6

Agricultural Chemicals Approval Scheme —Approved Products for Farmers and Growers, 174, 190, 226

Agriculture and Horticulture Act, 1964, 228

Alkethene pipes, 139

Aluminium alloy for greenhouse structure, 22

Ammonium nitrate, 65

Ammonium phosphate, 65

Analysis, soil and tissue, 70–3, 120

Aphids, 189, 193, 226

Auchincruive, research station at, 86, 96

Bacteria, beneficial, 56f

Base fertilisers, analysis of, 64; application of, 120–3, 222

Benches, 42–3; capillary, 45

Benomyl, 178f, 181, 226

'Black bottoms', 52, 58, 107, 197

Blight forecast, BBC, 183; *see also* Potato blight

Blossom end rot, *see* Black bottoms

Blotchy ripening, 197

Bolsters, 107, 128

Bone meal, 65

Border culture, 105–6, 114–24 (preplanting), 151 (first waterings), 158–66 (general), 171–2 (end of season)

Boron, 63, 195

Botrytis, 156, 171, 180–1, 191–2, 226

Broadbent, Dr L., and virus testing, 218

Bronzing, 197

Brown root rot, 177, 191, 226

Bubble house, 25, *50*

Buck-eye rot, 183, 192

Bushel, 70, 84, 85

Bush tomatoes, 224–5; *see also* Variety list

Calcium, 62, 66, 70, 74f, 196

Capillary benches, 45

Capillary watering system, 81

Captan, 180f, 226

Carbon, 56; bisulphide, 204, 206

Carbon dioxide, and nitrogen, 58; and photosynthesis, 48, 158–9; and straw bale culture, 131; enrichment technique, 41–2

Chemicals for the Gardener, 174

Chemicals, list of, 226–7; warning on, 174

Chemical sprays, 182

Chemical sterilisation, 123, 125, 204–5, 221, *168* (crop)

Cheshunt Compound, 175, 191

Chlorophyll, 48, 61

Chloropicrin, 205f, 226

233

Chlorosis, 61, 63
Circulating pump, 33f, 38
Cladosporium, 182, 192, 226; varieties resistant to, 210–17
Climate and tomato culture, 12f
Cloches, 11, 224–5
Collar rot, 176
Compost, 78–86; for potting, 94; municipal, garden and brewery, 118–19
Concrete barriers, problems with, 116
Conductivity factor (Cf), *see* Salt content
Container culture, 107, 115, 128–31 (pre-planting), 144 (planting), 146 (root development), 152 (first watering), 166, 169 (general)
Containers, 93–4 (potting), 107, 125, 126–7, 128–31 (planting), 222 (outdoors)
Convector heaters, 32
Corky root rot, 177, 191, 226
Creosote, for botrytis, 181
Cresylic acid, 172, 187, 204, 206, 226
Cropping pattern, varieties and, 210–17
Crop programming, 88–91
Cucumber mosaic, 185, 192
Cultural methods, balance sheet, 105–7; choice of, 113–14; selection chart, 115
Cuttings, propagation from, 87

Damping-down, 53, 109, 111, 146
Damping-off, 175, 191
D-D, 187, 204, 206, 226
Defoliation, 14, 156, 182, 224 (outdoors)
Didymella, 180, 191, 226
Die-back of roots, 145
Digging, 117
Dilutors, 41
Diseases, 14, 105f, 174–86, 191–2, 226; and humidity, 171; and tempera-

ture variations, 148
Dithane, 178, 205f
Double streak, 184, 192
Double truss cropping, 134–5
Drainage, for greenhouse site, 23; in borders, 114, 116; in ring culture, 125; outdoors, 221; polythene problems, 128ff, 131
Dried blood, 65, 77, 100
Dry heat sterilisation, 203, 206
Dry set, 198
Dutch light greenhouse, 23, *49*

Electric controls, 35, 38
Electric Growing, 44
Electric heaters, 32, 34f, 36 (cost/efficiency table)
Electricity supplies, 23, 27
Elements essential to plant growth, 55–66
Environmental control, 37, 39
Epsom salt, 66
Establishment, ideal conditions for, 146; outdoors, 223

'Fade-out' plants, 171
Fan heaters, 32, 35, 36 (cost/efficiency table)
Fans, *see* Ventilation
Farmyard manure, 60, 118, 124, 166, 221 (outdoors)
Feeding, amounts required, 74–8; and young plants, 99; end of season, 162–3, 171; general procedure, 159–60, 161–5; osmotic, 147, 162; outdoors, 223; *see also* Fertilisers, Nutrients
Fertilisation, 52–5
Fertilisers, and limited growing media, 170; and osmosis, 52; and potting media, 94; and young plants, 100–1; base, 63–6, 74–6; excess of, 100; for outdoor plants, 223; for straw bale fermentation, 132, 134;

general procedures, 161–6; laboratory analysis of, 73; liquid, 41, 77–8; *see also* Feeding, Foliar feedfeeding, Nutrients
Flooding, 58, 108–9, 121
Flower disorders, 198–9
Flower, tomato, 53f
Flower trusses, at planting, 141; early formation of, 88, 158
Foliage, removal of, 14, 156, 224
Foliar feeding, 99, 165, 187
F1 hybrids, 207, 210–12
Foot rot, 176, 191
Formaldehyde, 94, 126, 172, 204, 206
Fruit picking, 88–91, 170, 172
Fruit splitting, 197
Fumigation, 172
Fungal diseases, 175–83, 191–2
Fungi, and checked roots, 145
Fungicides, 175–83
Fusarium wilt, 179, 191; varieties resistant to, 211–13

Germinating cabinets, 93
Germination, 87–8, 92–3
Glasshouse Crops Research Institute, 218
Glasshouse research station, Ayrshire, 96
Glass in greenhouse structure, 11, 15f, 24f, 28f; cleaning of, 87 (*see also* Hygiene)
Grading, 170, 228–9
Grafted tomatoes, preparations for planting, 123; culture of, 169
Grafting, against eelworm, 187; against wilts, 179; procedure, 101–3, 167; sowing for, 92
Greenback, 197; varieties free from, 210–17
Greenhouse, 11–46 (*see* chapter breakdown, 5); cleansing, 87, 172; paths in, 137–8; *49, 50*

Growing methods, a balance sheet, 105–7
Growing rooms, 43–4
Growth factors, requiring attention, 158–61

Heating systems, 28–35, 36, 67
Heat input, calculation of, 28–32
Heat loss, 11, 15f, 28–32, 33, 230
Heat sterilisation, 201–3
Hedges for shelter, 20–1, 26
Hoddesdon method (sterilisation), 202
Hoof and horn meal, 65, 84
Hormone damage, 192, 199
Hormone setting liquid, 161
Hosepipe watering, 40
Hot water pipes, 31, 33–4, 38
Humidity, and botrytis, 181; and cloche culture, 224; and establishment, 146f; avoidance of, 13; late season, 171
Hydroponics, 83
Hygiene, 41, 87f, 94, 109, 180f, 185

Iron, 62–3, 195

John Innes preparations, base, 64, 122; composts, 81, 84, 126–7, 175, 221

Layering, 142
Leaf curling, 198
Leafhopper, 226
Leafminer, 189–90, 193, 226
Leaf mould disease, *see* Cladosporium
Leaves, removal of, 14, 156, 182, 224
Levington compost, 84, 130
Lighting, supplementary, 44–5
Light intensity, 9, 13; effect of, 160–1
Light levels, and growth, 97; and ripening of bottom trusses, 90
Lime, 59, 62, 66, 70, 120, 221–2 (outdoors)
Liquid feeding, quantities for, 77–8

Loam, 78–9, 82–3
Lycopersicon esculentum, 9

Magnesium, 60, 66, 74f, 122, 123, 195
Manganese, 63, 195
Manures, 60, 118
Metal structure greenhouse, 22
Metham-sodium, 124, 199, 205–6, 226
Methyl bromide, 205f
Mice, 190, 194
Millipedes, 190, 194
Mineral disorders, 194–6
Mobile greenhouses, 23–4
Mulching, 166, 178f, 187, 223f (outdoors)

Necrosis, leaf, 61
Netherlands, crops in sand, 116
Nitram, 65
Nitra-shell, 65
Nitrate of potash, 65
Nitro-chalk, 65, 186, 192
Nitrogen, 57–8, 60, 64, 65f, 121; availability analysis, 70–1; excess and deficiency of, 107, 159; fertilisers, 64f, 121–3; high nitrogen feeds, 64, 122–3, 162–3, 165, 171; nitrogen/potash ratio, 122, 147, 164
Nitro, 26, 65
Nutrients, application of, 120–3; availability of, 74; calculating quantities of, 77–8; excess and deficiency of, 159–60; requirements, 74–6; *see also* Feeding, Fertilisers

Oedema, 198
Oil-fired heating, 35, 36 (cost/efficiency)
Oil heaters, 34
Osmosis, 51–2, 54; osmotic equilibrium, 71; feeding, 147, 162

Outdoor culture, 220–5; varieties for, 211–17

Parasites, 145, 176
Pasteurisation, *see* Sterilisation
Peat, and outdoor tomatoes, 221f; as growing medium, 60, 79–80, 83, 86, 129–30, 144; as organic conditioner, 60, 118; for bench covering, 43; for mulching, 60, 166, 223; pots, 94
Peat mattress culture, 107, 130, 152
pC figures, *see* Salt content
Pelleted seed, 92, 216
Perlite, 81
Pests, 73, 105, 186–90, 193–4, 226–7; in garden compost, 119; in manure, 118; straw bale culture and, 107
pH figures, 62, 69f, 120 (for tomatoes), *see also* Calcium, Lime
Phenols, 204–5
Phloem, 52, 54
Phosphorus, 58–9, 60, 70; excess and deficiency of, 160; tomato requirements of, 74f
Photosynthesis, and carbon dioxide, 48, 158–9; and other nutrients, 57, 59, 61ff; germination and, 88; spacing and, 98
Physiological disorders, 196–9
Physiology of tomato plant, 47–55
Pipes, soil-warming, 33–4, 139
Planting, distances, 136–8; layering, at, 142; method of, 142; soil warmth and, 138–9, 140, 142; stage of plant at, 141; times, 88–91, 139–40, 222 (outdoors)
Plastic, for greenhouses, 24–5, 50; pots, 94; waste, as growing medium, 82; *see also* Polythene
Pollination, 52–5, 109
Pollution, 13
Polystyrene, as draining material, 128; as growing medium, 81f

Polythene, as greenhouse lining, 45–6, 175; barriers, 116, 125; cloches, 224; curtains, 46, 100; heat loss with, 11, 16

Polythene cultural methods, bag, 107, 112, 128; bucket, 107, 130–1; trough, 107, 115, 129, 152; trench, 131, 152, 221 (outdoors)

Polyurethene foam, as growing medium, 82

Potassium, potash, 59f, 61, 69f, 84; application of, 121–3; excess and deficiency of, 160; high potash feeds, 122–3, 162–3, 171; medium potash feeds, 122–3, 222f

Potato blight, 183, 192, 226

Potato eelworm, 105, 116, 186–7, 193, 221, 226; count, 72

Pots, *see* Containers

Potting-up, containers for, 93–4; programme for, 88–91, 95

Poultry manure, 60

Programming crop, 88–91

Pruning, 155–7

Red spider mite, 188, 193, 226

Removal of plants, 172

Requirements, basic, for tomato culture, 11–14

Respiration, 47–8

Ring culture, 106, 112, 115, 124–7 (pre-planting), 143 (planting), 151 (first waterings), 166, 169 (watering and feeding), 68

Rogue plants, 93

Root knot eelworm, 187–8, 193, 226

Root rot, 176–8, 191

Roots: adventitious, 178, 191; deep, 117; development of, 145–6; examination of, 113, 124, 172; two-zone system, 124, 127

Root stock, 101–3

Rotary cultivation, 117

Rotary drum sterilisers, 203, *168*

Salt concentration meter, 71

Salt content, soluble (pC, Cf), 71; and container culture, 107; and nutrition, 159, 162; and polythene bag culture, 129; and potting media, 94; and root damage, 145; checking, 109, 119, 121, 124

Sand as growing medium, 80–1, 83, 116

Sclerotine stem rot, 183, 192

Seaweed, 60

Seed, pelleted, 92, 216; saving, 218–19; *see also* Sowing

Severe streak, 184, 192

Sewage, sludge, 60

Shading, 178, 225

Shelter, provision of, 20–1

Shelters, portable, 220, 223

Side-shoot removal, 155–6, 223f (outdoors)

Silvering, 198

Single truss cropping, 134–5

Site, for greenhouse, 17–20, 25; for outdoor tomatoes, 220

Sodium chlorate, 117

Soil: analysis, 69–73, 120f; formation of, 55–6; preparation of, 116–17, 221 (outdoors); sickness, 200; state of, at planting, 142–3; *see also* Sterilisation

Soil blocks, 93–4, 222

Soil-less blocks, 94

Soil-less culture, 74, 81–5, 106–7, 124–34; and 'feather roots', 147

Soil-warming, 33–4, 42–3, 138–9, 140, 142

Solid fuel heating, 35, 36 (cost/efficiency)

Soluble salt content, *see* Salt content

Sowing, 91–2; times for, 88–92, 222

Spacing, 136–8; outdoors, 222; times for, 88–91, 98–9

Spectrochemical analysis, 69, 71–2

Spotted wilt, 184, 192

Spraying, outdoor, 223; *see also* Watering
Spraylines, 40, *68*
Springtails, 188, 193, 226
Sprinklers, 40
Steaming, 202–3, 206
Stem rots, 180–1, 183, 191–2, 226
Sterilisation, 200–6, *168*; and borders, 105, 116, 119, 122ff; and eelworm, 187; and nitrogen flush, 57–8; and symphalids, 188; chemical, 123, 125, 204–5; of loam, 79
Stopping, 157, 223
Storage heaters, 35
Straw bale culture, 107, 112, 115, 131–4 (pre-planting); 138, 144 (planting), 151 (first waterings), 169 (general), *150*
Suckers, 156
Sulphur, 63–4; flowers of, for botrytis, 181; for red spider mite, 188; for white fly, 189
Sun, and greenhouse, 16–19; shut-off, 13, 20, 25f
Support, 153–5, 223 (outdoor); in plastic structures, 25; of young plants, 99; on straw bales, 134
Symphalids, 188, 193, 226
Systemic fungicides, 176

Temperatures: and first trusses, 96, *149*; and fruit quality, 140–1; and germination, 93; and heating systems, 32f; and rate of growth, 161; conversion table for, 231; table of recommended, 97–8, 148; variations, effects of, 148
Testing kits, soil, 69
Thermometer, max and min, 98; soil, 142
Thermostat control, 35, 38, 43
Thrips, 189, 193, 227
Toe rot, 177, 191
Tomato mosaic virus, TMV, 184,

192; varieties resistant to, 210–17
Tomato moth, 190, 193, 227
Tomato plant, diagram of, 54; history of, 9
Training, 133, 134, 154–5, 223–5 (outdoors)
Translocation, 52, 54
Transpiration, 48–9, 54
Trench system, 131; and first waterings, 152; outdoors, 221
Trickle irrigation, 40, *68*; and root formation, 147
Tri-sodium phosphate, 88, 95, 185
Trough culture, 107, 129, 138, 144, 152 (first waterings)
Trusses, *see* flower trusses
Truss limitation, 134–5, 223f (outdoors)
Turbulence, avoidance of, 20

Urea, 65; formaldehyde, 65, 82

Varieties, 207–9 (general classification), 209–17 (list of), 218–19 (notes on)
Ventilation, 22, 25–6; and damping-off, 175; and disease, 181f; for plant establishment, 146; for young plants, 100; systems, 35, 37–8, *67*
Vermiculite, 81, 130f
Verticillium wilt, 179, 191; varieties resistant to, 210–13
Virus infection, 184–6, 192; and cuttings, 87; and seedlings, 88; and side-shoot removal, 156; and young plants, 101; in border culture, 106, 116
Virus testing, 218

Waste paper, as growing medium, 82
Water, bulk effect on radiation, 13; supply to greenhouse, 23, 27, 39–40; uptake rate, 51
Watering, 108–9, 119, 121 (pre-

planting); 142–4 (planting); 111–12 (requirements table), 146–7 (establishment), 151–2 (general), 171 (end of season), 223 (outdoors); excess, 58, 100; rootball, 151f; systems, 40–1; *see also* Flooding

'Weaning', 147

Weedkillers, action of, 52; warnings on, 117–18, 192, 199

Weeds, control of, 23, 27, 117–18; in border culture, 105; in FYM, 118

Whalehide pots, for ring culture, 125, 126, *68*; on border soil, 130

White fly, 189, 193, 227

Wilting, 101, 187

Wilts, 179, 191

Wind, exposure to, 11, 20–1, 25, 220 (outdoors)

Wireworms, 190, 194

Woodlice, 190, 193, 227

Xylem, 48, 51, 54

Yields, 172

Young plants, guide to raising (table), 100–1